CW01647071

C

Consulting Ltd

Assembly Line Planning and Control

Nick T. Thomopoulos

Assembly Line Planning and Control

Springer

Nick T. Thomopoulos
Stuart School of Business
Illinois Institute of Technology
Chicago, IL
USA

ISBN 978-3-319-01398-5 ISBN 978-3-319-01399-2 (eBook)
DOI 10.1007/978-3-319-01399-2
Springer Cham Heidelberg New York Dordrecht London

Library of Congress Control Number: 2013945141

Printed on acid-free paper

Springer is part of Springer Science+Business Media (www.springer.com)

For my wife, my children and my grandchildren

Preface

I was fortunate to have a rich and diverse career in industry and academia. This included working at International Harvester as supervisor of operations research in the corporate headquarters; at IIT Research Institute (IITRI) as a Senior Scientist, serving in the Advance Assembly Methods program with worldwide applications; as a Professor in the Industrial Engineering Department at the Illinois Institute of Technology (IIT); in the Stuart School of Business at IIT; and many years of consulting assignments with industry and government throughout the world. At IIT, I was lucky to be assigned a broad array of courses, gaining a wide breadth with the variety of topics, and with the added knowledge I acquired with every repeat of the course. I was also privileged to serve as the Advisor to many bright Ph.D. students as they carried on their dissertation research. Bits of knowledge from various courses and research helped me in the classroom, and also in my consulting assignments. I used my industry knowledge in classroom lectures so the students could see how some of the textbook methodologies actually are applied in industry. At the same time, the knowledge I gained from the classroom helped me to formulate and develop solutions to industry assembly line applications as they unfolded. This variety of experience allowed me to view how assembly systems are used in industry. This book is based on this total experience and also includes the quantitative methods that I found doable and useful.

Thanks especially to my wife, Elaine Thomopoulos, who encouraged me to write this book, and who gave consultation whenever needed. Thanks also to many people who have helped and inspired me over the years. I can name only a few here. These are: Robert Battaglia, Corning Glass; Sven Berg, L. M. Ericsson; Harry Bock, Florsheim Shoe Company; Fred Bock, IITRI; Randy Braun, Komatsu-Dresser; Toemchai Bunnag, PTT Public; John Cada, Florsheim Shoe Company; Michel Chaussumien, Citroen; Janis Church, IITRI; Charles Clerval, Citroen; John DeMotts, International Harvester; Ralph Edgington, NCR; Jim Gleason, International Harvester; Scott Haligas, Florsheim Shoe Company; Al Hawkes, IITRI; Joe Hoffman, Springfield Remanufacturing Corporation; Linkert Horn, IBM; Maurice Kilbridge, Harvard University; Jack Kornfield, IITRI; Melvin Lehman, Loyola University; Ronald Lodewyk, California State University, Turlock; Nico Mantelli, Olivetti; Craig Marecek, Komatsu-Dresser; Joe Moore, IITRI; Chad Myers, Springfield Remanufacturing Corporation; Jerry Novak, IITRI; Paul O'Donnell, Westinghouse; Dino Olivetti, Olivetti; Stylianos Papaioannou,

Lapin House; Ernest Pirone, IITRI; Ted Prenting, Marist College; Serge Rhandhava, IITRI; Harry Rindeskar, ASEA; Walter Ryder, University of Southern California; Ryan Stack, Springfield Remanufacturing Corporation; John Storto, IITRI; Marvin Sussman, International Harvester; Ola Tannous, Electro Motive Diesel; Tor Wadell, ASEA; Leon Wester, IITRI; and Charles Wright, Navistar.

Nick T. Thomopoulos

Contents

Notations

a	Learning time for first repetition
A(r)	Average learning time for repetition r [Ar = A(r)]
b	Bill-of-material pieces per unit
b	Learning rate coefficient
b(h)	Bill-of-material quantity for part h
b_h	Bill-of-material pieces per unit. Note, [b(h) = b_h = b]
c	Operation time
$\bar{c}$	Average operation time
c_{ij}	Operation time at station i for model j
d	Balance delay
ds	Days-supply
E	Line efficiency
e	Work element
e(h)	Prime element of part h
f	Features
g	Labor group
h	Parts
i	Stations or operators
j	Models or jobs
j*	Set of models in similarity index and learning
k	Learning multiplier
k	Options of feature f
k(j,f)	Option k of feature f for job j
L	Lead time
M	Multiple quantity
n	Number of stations
N	Shift schedule quantity
Nd	Number of days
Ne	Number of elements
Nf	Number of features
n_{fk}	Shift number of options k for feature f
N(f,k)	Shift number of options k for feature f. Note, [n_{fk} = N(f,k)]
N_{fk}	Number of options k for feature f
Nh	Number of parts

Ni	Number of stations
Nj	Number of models
N_j	Shift schedule for model j
Nk_f	Number of options for feature f
oh	On-hand inventory [oh = OH]
ohoo	On-hand plus on-order inventory
oo	On-order inventory [oo = OO]
OL	Order level
OP	Order point
P_{buy}	Parameter of days-supply for buy quantity
P_{fk}	Probability of option k for feature f
P_{ss}	Parameter of days-supply for safety stock
P(f,k)	Probability of option k for feature f [$P(f,k) = P_{fk}$]
q	Buy quantity
Q	Buy quantity in multiple
Q	Mixed model learning coefficient
R	Learning rate
r	Repetitions in learning
R_h	Shift requirement of part
R_{hij}	Shift requirement of part h at station i for model j
r_o	Learning limit
S(j*)	Similarity index for model set j*
T	Shift time
t	Work element time
t(e)	Work element time for element e
T(r)	Cumulative learning time for repetition r
t(r)	Learning time for repetition r
T_e	Shift time for element e
t_e	Time per element e [$t_e = t = t(e)$]
T_{ej}	Shift time for element e on model j
t_{ej}	Time for element e for model j
T_i	Shift time for station i
T_j	Shift time for model j
T_j	Unit time for model j
$\bar{T}j$	Weighted average unit time for all models
U(j*)	Utilization index for model set j*
u_{ej}	Usage index of element e on model j
u_{hj}	Usage index of part h on model j
v	Conveyor speed
w	Unit space
Σt_e	Unit time (standard work time for a unit)
Σt_{ej}	Unit time for model j

Chapter 1
Introduction

The advent of the assembly line is often credited as one of the most significant developments in the modern world. In 1776, Adam Smith had the foresight to describe the value of adopting the division of labor in his book *The Wealth of Nations*. Shortly later, in 1797, Eli Whitney showed the importance of using interchangeable parts on a mechanized assembly line to manufacture muskets for the U.S. government. He created standard parts that were used in the assembly process, and that could also be used to replace any damaged part that might be needed subsequently. It wasn't until the beginning of 1910 when Henry Ford employed assembly lines with conveyor belts to mass-produce the Model T automobile for the Ford Motor Company. With this achievement, Ford essentially paved the way to a whole new era in manufacturing. Today assembly lines are in use globally in all types of industries.

This book delineates the various quantitative methods that are used in assembly lines and demonstrates how they can be applied. The book includes applications in single model lines, make-to-stock mixed model lines, make-to-order mixed model lines, postponement lines, and one-station assembly. The book shows how to select the quantity of units to schedule for a shift duration, compute the number of operators needed on a line, set the conveyor speed, coordinate the main line with sub assembly lines, assign the work elements to the operators on the line, sequence the models down the line, sequence the jobs down the line, calculate the part and component requirements for a line and for each station, determine the replenish needs of the parts and components from the suppliers, compute a similarity measure between the models being produced, use learning curves to estimate time and costs of assembly, and measure the efficiency of the line.

Review of the Chapters

The quantitative methods presented in Chap. 3 through 12 are described with examples throughout. The examples pertain to the assembly of one or more models on the line. For brevity sake, the number of work elements used in these examples

N. T. Thomopoulos, *Assembly Line Planning and Control*,
DOI: 10.1007/978-3-319-01399-2_1,

is smaller than most real situations, but adequate in size to allow the reader to follow the calculations, and sufficiently large enough to illustrate the methodology.

Chapter 2, **Assembly Systems** This chapter is partitioned into two sections: History of Assembly Lines and Type of Assembly Lines. The history begins in 200 BC in China where 8,000 sculptures known as the Terracotta Army were produced in an assembly type manner. In the 16th Century, the Venetian Arsenal built standardized parts to equip its ships (galleys) using assembly methods. Oliver Evans of Delaware, in 1785, applied assembly methods in a flour mill; Eli Whitney, in 1797, used assembly lines to build muskets for the U.S. government. Assembly systems were in use in England at the Portsmouth Block Mill in 1800, and at the Bridgewater Foundry in 1836. Assembly systems were also in use in the U.S. during the 19th century, particularly in the armories; and in 1867, in the meatpacking industry in Chicago. In 1901, Ransom Olds was the first to apply assembly methods to automobiles, increasing the output from one unit to five units a day. Henry Ford designed and built the first successful automotive assembly line. His factories produced hundreds of Model T Fords each day. Ford's innovative methods paved the way for the use of assembly lines all over the world. The chapter also describes the various types of assembly lines: single model assembly, batch assembly, mixed model assembly for make-to-stock, mixed model assembly for make-to-order, postponement assembly, one station assembly, disassembly and robotic assembly.

Chapter 3, **Some Fundamentals** This chapter describes the data, computations and decisions needed to control an assembly line. The key data includes the work elements, the element times, the predecessor elements, the unit time, shift time and the shift schedule quantity. The data is used to compute the number of operators needed on the line, the cycle time, average operator time, balance delay and the efficiency ratio. The management can now assign the work elements to the operators so that the operator times are evenly distributed and all precedence constraints are satisfied accordingly. This step is called line balancing. Next, the operator times are measured along with the effective cycle time and effective balance delay. The bill-of-material of the product(s) now enters as data to determine the part and component requirements needed for a shift duration at every station along the line. The chapter also describes the relation between the main line, feeder lines, subassembly lines and labor groups, as well as the use of parallel stations and parallel lines.

Chapter 4, **Assembly Planning** Through examples of single model and mixed model assembly lines, this chapter shows how to make the best decisions in order to meet the production plans specified while attaining efficiency in the assembly operation for future time periods. An example of a single model line shows the data and computations that allow management to select the production schedule for the line. A range of shift schedule quantities is considered and with each quantity, the number of operators needed on the line and the associated efficiency measures are calculated. The first example is a single model assembly situation with one labor group, and the next contains two labor groups. The computations give the speed of the conveyor system and also the length requirements of the assembly line. When more than one labor group is needed, the cycle times and conveyor speeds need to be coordinated so the units flow appropriately from one labor group to the next. Examples of mixed model make-to-stock assembly with one labor group and with five labor groups demonstrate the data and computations required to determine the number of operators to assign in each labor group, and how to measure the operation efficiencies.

Chapter 5, **Inventory Replenishments** An important step in planning the assembly of a product is ensuring all of the parts on the bill-of-material are available at the station locations at the start of the assembly process. This chapter shows how to determine the part requirements for single model lines, for mixed model make-to-stock lines, and for the mixed model make-to-order lines. Sometimes the station storage space is limited and therefore just the right quantity of stock is necessary. All of the inventory can be stocked at each station location at the start of a shift, or can be stocked two or more times during the shift in a just-in-time manner. In either event, the requirements are stocked prior to their need on the line. The chapter also describes how to control the daily part replenishments coming from the suppliers. When the assembly line for a product is run day after day, the part requirements over the planning horizon are computed for forthcoming shifts. The future part replenishments use the projected daily shift requirements to determine when, and how much, new replenish stock is needed. The replenish projections over the planning horizon for each part is useful information to the part suppliers allowing them to plan their production activities accordingly.

Chapter 6, **Single Model Assembly** This chapter concerns a plant that dedicates a line to produce a product that has no variation. This is called single model assembly. The planning methods that take place for this type of line are described in the context of an example. The example begins with the work elements, the element times, the predecessor elements, and the corresponding precedence diagram. The shift schedule quantity and shift time are needed to determine the number of operators to have on the line. The example shows how to assign the work elements to stations (line balancing) in order to obtain an even work load per station, as well as attain compliance with the precedence constraints. The example continues by showing how to measure the balance delay and efficiency ratio for the line. The bill-of-material data is used to identify the relation of parts to the work elements. This data allows the management to compute the requirements of parts for the shift schedule and for each station. The example also shows how the part requirements over the shift are replenished from the supplier. The replenishments could occur one time for the entire shift, or two or more times over the shift in the spirit of just-in-time deliveries.

Chapter 7, **Mixed Model Make-to-Stock Assembly** Mixed model make-to-stock assembly occurs when one line has two or more models in process at the same time. This chapter describes the planning methods that take place to control the operation of the line. The methods are presented by an example or four models. The example begins with a listing of the work elements, the element times, the predecessor elements, and the element relation with the models, called the usage index. The shift production time and the shift schedule for each model are also needed here. Next the stations are assigned the work elements where the assigned times over the shift are evenly distributed (line balancing). Each day, the sequence of the units down the line is generated in a way where the flow of work is as smooth as possible with minimum idleness and congestion at the stations. A make-to-stock sequencing algorithm (MSSA) demonstrates how the method works. Finally, the bill-of-material data is called to calculate the part requirements for the each shift and for every station.

Chapter 8, **Mixed Model Make-to-Order Assembly** Make to order assembly occurs when the customers specify the features and options for each unit they buy. The methods to control the operation of this type of assembly line are described by an example. The example begins with the work elements, element times, predecessor elements and any associated features and options of the elements. The probability of options by feature is used to project the number of options and features by shift over the planning horizon. The shift time and the number of units to build over the shift are then used to estimate the element times over the shift. With this information, the assignment of elements to the stations (line balancing) is carried out. An order board contains all the current customer orders, called jobs, with the exact feature and option combinations, as well as the due dates. The management selects the jobs for a forthcoming daily shift. The next decision is how to sequence the jobs down the line. A make-to-order sequencing algorithm (MOSA) is introduced and demonstrated with a shift schedule of fifty jobs. Sometimes, before the sequence date, a job that is scheduled on a sequence has to be removed and replaced by another job taken from a pool of candidate jobs. A make-to-order replacement algorithm (MORA) is presented to select the substitute job, from the pool of jobs and is demonstrated in the example. Finally, the bill-of-material for the parts enters as data and is used to determine the part requirements for each shift and station.

Chapter 9, **Postponement Assembly** Postponement is a strategy that can be applied to products such as computers, trucks, automobiles and farm tractors that are offered with a variety of features and options. In the assembly process, the units are built without the variety of features and options. The assembly is like a single model line and the output units are stocked in a warehouse facility. When the customer orders come in with the exact feature and option combination, the final assembly takes place in the warehouse. This way, complicated make-to-order assembly is replaced with the simpler single model assembly. This strategy yields less inventory in the plant and reduces the lead time to customers. For convenience in this chapter, the strategy is called full postponement. Two alternative assembly strategies for this environment are demonstrated in comparison: no postponement and partial postponement.

Chapter 10, **One Station Assembly** One station assembly is described in the context of a shoe manufacturing plant where one worker is assigned a set of shoes by style and size to assemble all alone. The worker is given a batch of the items to produce, and is provided all the parts and components needed to complete the task. Multiple pairs are assigned to the worker; typically six to twelve pair at one time. The worker completes all the pairs in the batch prior to starting the next assignment. This is an example of one-station assembly. In other situations, the workers are assigned one unit at a time, as in engine assembly. This chapter describes some of the quantitative methods that are related to one station assembly. Sometimes the station operator requires a mold of some type to carry out his/her work. The mold is used in the production process and then can subsequently be used for another unit. The plant has an inventory of molds to allow the workers to carry out their assignments. The chapter shows how to determine the number of molds to have in the plant in order to yield a specified service level. The service level is the probability the mold will be available when needed by a worker.

Chapter 11, **Similarity Index** This chapter pertains to a mixed model assembly plant, and measures the similarity between the models based on the work elements. Two types of measures are developed, the utilization index and the similarity index. Both indices are measures of the similarity between the models. The latter index ranges from 0 to 1, where zero occurs when there is no similarity and one is when there is full similarity. The indices are developed from the work elements where some of the elements are common to all of the models, some are unique to a particular model and others are common to two or more models. Three scenarios are described: (1) where the elements and model usage are (0 or 1) and 1 indicates the element does apply with the model; (2) where the elements are (0 or t_e) and t_e is the common element time to all models where the element applies; and (3) where the elements are (0 and t_{ej}) and t_{ej} is the time for element e and model j. The indices can be measured for sets of two or more model combinations. Examples are given to illustrate how the similarity index may be used in assembly planning.

Chapter 12, **Learning Curves** Learning Curves can be used to estimate the time required to complete a selected number of units on an assembly line. The chapter shows how to apply learning curves for single model lines and for mixed model lines, and describes the learning rate, the learning coefficient, and the learning multiplier and how they are used to develop the learning curve. The learning curve and the unit standard time are combined to compute the learning limit for the product. The unit time is higher than the standard time for all assemblies prior to the learning limit, and for those after the learning limit, the assembly unit time is the same as the standard time. With all this information, the projected average time (and the cumulative total time) to complete a selected number of units can be computed. The chapter shows how to extend the method of learning curves for mixed model lines. Examples are given for a single model assembly line, for a 2-model assembly line, and for a 3-model assembly line. The method extends to an M-model assembly line.

Chapter 2
Assembly Systems

Introduction

An assembly line is a manufacturing process where the bill-of-material parts and components are attached one-by-one to a unit in a sequential way by a series of workers to create a finished good product. All of the tasks to fully produce the product are identified and as much as possible the task times are evenly assigned to the workers, whereupon the units are produced one after the other. This method of production has proven to be much more efficient than having a series of craftsmen entirely producing each finished good product one at a time. Adam Smith in the 1776 classic "The Wealth of Nations," introduced the term "division of labor". This involves the partition of a complex production process into one or a few simpler tasks, with each task assigned to a different worker. Smith gives an analysis of a pin factory where the time and physical movement of the workers were reduced throughout.

This chapter is partitioned into two sections: History of Assembly Lines and Type of Assembly Lines. The history begins in 200 BC in China where 8,000 sculptures known as the Terracotta Army were produced in an assembly type manner. In the sixteenth century, the Venetian Arsenal built standardized parts to equip its ships (galleys) using assembly methods. Oliver Evans of Delaware, in 1785, applied assembly methods in a flour mill; Eli Whitney, in 1797, used assembly lines to build muskets for the U.S. government. Assembly systems were in use in England at the Portsmouth Block Mill in 1800, and at the Bridgewater Foundry in 1836. Assembly systems were also in use in the USA during the nineteenth century, particularly in the armories; and in 1867, in the meatpacking industry in Chicago. In 1901, Ransom Olds is credited as the first to apply assembly methods to automobiles increasing his output from 1 to 5 units a day. Henry Ford designed and built the first successful automotive assembly line. His factories produced hundreds of Model T Fords each day. Ford's innovative methods paved the way for the use of assembly lines all over the world. The chapter also describes the various types of assembly lines: single model assembly,

N. T. Thomopoulos, *Assembly Line Planning and Control*,
DOI: 10.1007/978-3-319-01399-2_2,

batch assembly, mixed model assembly for make-to-stock, mixed model assembly for make-to-order, postponement assembly, one station assembly, disassembly, and robotic assembly.

History of Assembly Lines

Terracotta Army One of the earliest discoveries of an assembly line took place in China in 200 BC. In 1974, a collection of sculptures depicting the armies of the first emperor of China was discovered by a group of farmers. The discovery included 8,000 figures of warriors and chariots that were buried with the emperor in 210 BC. Historians state that the army figures were manufactured in workshops by laborers and craftsmen at the direction of the government. The parts of the figures (head, arms, legs, and torsos) were created separately and then assembled. Final touches were added to give different facial looks, and each workshop inscribed its name to the units they produced to ensure quality control. The process was like an assembly line production, where the individual parts were first formed, then fired, and later assembled.

Venetian Arsenal In the sixteenth century, the Venetian Arsenal employed 16,000 people who manufactured standardized parts of ships (sails, oars, wheel carriages, guns, rigs, ropes, munitions, etc.), in an assembly line manner. These parts were used to fully equip newly built galleys that were produced on a basis of almost one a day. A galley is a type of ship that is propelled by many rowers on both sides and was used for warfare, trade, and piracy.

Oliver Evans In 1785, Oliver Evans, in the state of Delaware, built the first automatic flour mill. The mill used a leather belt bucket elevator, screw conveyors, canvas belt conveyors, and other mechanical devices to completely automate the process of making flour. The innovation spread to other mills and breweries.

Eli Whitney In 1797, Eli Whitney used an assembly line to mass-produce muskets for the U.S. government. All the parts of the musket were produced in advance and with the same engineering tolerance so that each could be inserted onto any musket. In this way, the parts and components were common and made assembly possible. The common parts also were used subsequently to replace a musket part that had been damaged. Prior to Whitney, each musket was produced entirely by an individual craftsman. The parts and components were crafted to fit the individual musket. Because of this, the parts were not common and interchangeable from musket to musket.

A few years after in 1797, in firearm-manufacture, Whitney and other manufacturers began using machine tools and jigs to produce the parts and components. The equipment's ability to produce standard interchangeable parts with specified tolerance was used to replace the skill of the workers and allowed the hiring of less-skilled workers.

Portsmouth Block Mills In 1800, Samuel Boulton and James Watt, working at the Portsmouth Block Mills, in England, developed woodworking machinery for up-and-down saws and for circular saws. These were housed in a three-story woodworking building where the heavy products were transferred by flat belts running on pulleys. The machinery was used to cut timber into a variety of smaller parts (components for tables, benches, pumps, etc.), that were used in shipbuilding. Previously, these items were cut by hand. In 1802, The British Navy required 100,000 pulley blocks of various sizes, and was unsatisfied with the production of hand-made blocks. At Portsmouth Block Mills, Mac Isambard Bruner with Henry Maudslay and others designed 22 types of machine tools to make the parts for the blocks used by the Royal Navy, and successfully fulfilled the Navy's need. They are credited with one of the first linear and continuous assembly processing system.

Bridgewater Foundry James Nasmyth and Holbrook Faskell founded the Bridgewater Foundry in 1836 in England. Nasmyth designed and manufactured a large set of standardized machine tools (steam hammer, planers, shapers, pile drivers, hydraulic press, etc.), mainly for locomotive application. The firm is credited with using material handling methods in production that paved the evolution of the assembly line. Nasmyth arranged the factory in a line with a railway carrying the work from one building to another. Cranes were used to lift the heavy items. The production passed in a sequential way from erection of framework to final assembly.

American System During the nineteenth century, a wave of new manufacturing methods started in America allowing an upsurge in labor efficiency, particularly in the armories where products were produced for the U.S. government. This included increased use of interchangeable parts and mechanization of tools, fixtures, and jigs in the production process. The system applied specialized machinery instead of hand tools. This advance in manufacture is also referred as the armory practice since it was implemented mainly in the Federal armories.

The American system fostered continual innovations in machines, tools, fixtures, jigs, and part standards that could be carried out efficiently by semi-skilled workers. The workers were able to run the specialized machines that produced identical interchangeable parts that were made to the specified standards set by the engineers. All of this in minimum time compared to the individual craftsmen way.

In this era, the separation of manufacture of parts and components from assembly became possible. One set of workers could produce the parts and components, and in a separate facility, another set of workers could perform all of the assembly. Soon, this system became common in all the industries in the USA and around the world.

Meatpacking In 1867 in Chicago, the meatpacking industry created one of the first industrial assembly lines in the United States. The workers stood at their assigned stations as a pulley system brought the meat to their station allowing the worker to complete the task assigned to the station. This essentially is a disassembly operation.

Pre 1900 Prior to 1900, most manufactured products continued to come from the craftsmen who worked individually. The different parts were crafted one-by-one with files, knives, and other tools in a trial and error manner until they fit together. Then, the parts were assembled to produce the finished good item. In England and in New England, the manufacturing way slowly was changing. The advent of jigs, fixtures, machine tools, lathes, and planers paved the way to interchangeable parts and assembly applications. Different forms of assembly lines started to pop up in a variety of industries (textiles, clocks, horse drawn vehicles, railroad cars, sewing machines, and bicycles).

Ransom Olds In 1901, Ransom Olds developed the first automotive assembly line for the Olds Motor Vehicle plant in Detroit, Michigan, where the vehicle was the Oldsmobile Curved Dash. The new approach to putting together automobiles enabled Olds to increase the annual factory output from 425 cars in 1901 to 2500 in 1902. Ford improved Olds idea by installing conveyor belts. Some credit Olds as "the father of automotive assembly line", and Ford as "the father of automotive mass production." Although history cites Olds as the first user of automobile assembly lines, Henry Ford is recognized as the initial pioneer since he took the concept and perfected it for all generations.

Henry Ford Henry Ford's early career was a Chief Engineer in Thomas Edison's electrical lighting plant in Detroit, Michigan. His interest, however, was internal combustion engines, and had a desire to develop a vehicle that would be driven by one. In 1902, after Ford left Edison, he founded the Ford Motor Company. He started to build vehicles of various models, but was hampered because he had little cash available. He began to offer dealer franchises that required the dealers to pay for the vehicles upon delivery to them instead of after they were sold. With this added infusion of funds, he had capital to advance his manufacturing facility and continue researching to improve the model cars. He produced various models, including some for luxury and others for racing.

His vision was to build an auto that the common workingman could afford. In 1908, he introduced the Model T that would soon become the auto he was seeking. Up to that time, the vehicles were custom-built one at a time in small quantities. In 1909, he started a facility to apply assembly methods. Little by little, improvements took place. He installed moving belts so that the workers could remain at one location and do their one task efficiently and in minimum time, rather than assigned a variety of tasks. With the efficiency in production, the line soon was producing 1,000 vehicles per day. Ford was then able to lower the cost of the vehicle to $290, and this was in the range of the common man. By 1915, he produced almost half of the world's automobiles, and by 1923, the production rose to 1,800,000 per year. With all the efficiency in production, the worker's skill level was deteriorating. The assigned task per worker was not challenging and became boring. Some workers left the plant for other jobs more stimulating.

Ford recognized the problem and proved innovative in finance as well. In 1914, he offered his workers a wage of $5.00 per day; that was well ahead of the typical workers wage. Doing this, he gained many new workers. This high wage also allowed the workers to afford the vehicles they were producing; and as a result, his sales went up tremendously as a result. By 1927, over 15,000,000 Model T's were sold. With the proliferation in vehicles, the state and city governments required new roads and networks where vehicles could travel. This also included the need for a series of petrol stations and traffic controls. A whole new industry of tourism became fashionable. People could now visit places that previously were out of reach.

Type of Assembly Systems

Single Model Assembly A single model assembly is where the line is dedicated to one model. For example, a washing machine manufacturer produces three models (A, B, and C). The models are assigned to individual lines, and model A is placed on a line that runs on one shift per day of 450 min. Each unit of the model is the same with no variations. The schedule over the planning horizon calls for 200 units per day. Because A is the only model on the line, the line is called a single model line. A goal of the management is to assign the work elements associated with the model to the station operators as evenly as possible in way where the workload can produce the 200 units over the shift.

Batch Assembly Consider again the manufacturer with three modes (A, B, and C) and one assembly line. In batch assembly, the models are assigned to the line in pre-assigned sizes, where model A is run until its inventory is at a specified level, and in the same way, then model B is run and finally model C. The cycle continues where the models are assembled in batch sizes that satisfy the warehouse inventory levels allowing sufficient stock to meet the oncoming customer demands. With each change in model on the line, the station operator's workload needs to be changed accordingly. Essentially, each model is run as a single model line.

Mixed Model Assembly for Make-to-Stock Mixed model assembly occurs when more than one model of a product is assigned to the same assembly line at the same time. An example is the washing machine manufacturer with models A, B, and C, where all three models are assigned to a line. Over the planning horizon, the daily shift schedule calls for 100 units of model A, 70 of B, and 30 of C. The models are different, but the units of each model have no variation. The two main tasks for this type of line are line balancing and sequencing. Line balancing is the process of assigning the work elements to the station operations in a way where each operator's workload is as even over the shift as possible. Sequencing is the arrangement of sending the units down the line in a way where the workflow minimizes lapses of idleness and congestion over the shift and at all of the stations.

Mixed Model Assembly for Make-to-Order Mixed model assembly for a make-to-order manufacturer occurs when the manufacturer's product is offered with a series of features and options. Each customer order specifies the option for each feature of the product. This way, every customer order is unique and often no two orders are the same. The customer order is called a job. Typical of this type of manufacture is truck assembly, where the customers have individual needs on their vehicles. The assembly process first requires the task of assigning the workload to the station operators where the shift workload is as even as possible. This is difficult, since each day, the units coming down the line are different. Projection of features and options over the planning horizon are needed in assigning the workload to the individual stations. The management next has to determine which jobs to assign for assembly in a shift. Each job has a due date and a bill-of-material that is unique to the job. When all the parts and components for the job are available and the due date is proper, the job is scheduled for assembly on an upcoming shift. The management must then determine a sequence of the jobs for each shift that allows a smooth flow of the workload with the least amount of idle time and congestion time among all of the stations on the line.

Postponement Assembly Postponement assembly is a supply chain strategy that could apply to manufacturers that offer products that have a series of features and options, where the customer orders are for a particular combination of the options. Postponement is where the assembly line partially completes the products so that they can be fully completed at the later time prior to delivery to the customer. An application are college pennants that are produced and stored without any color. When the customer orders come in, the name and colors are inserted accordingly. Another situation is farm tractor assembly where the units are produced and stored in a stripped down manner awaiting customer orders with the specific features. As each customer order arrives, the features are inserted onto one of the stripped down tractors.

One Station Assembly One station assembly happens when all the assembly work to complete a unit is assigned to one person in one station. This could be an engine assembly in a small shop that builds specialty engines. Another example is a shoe manufacturer that assigns 12 pair of shoes from a particular style and a combination of sizes to a worker. At the outset, all of the material to complete the 12 pair is delivered to the worker's station. The worker may spend one or more days on completing the assigned pairs.

Disassembly A big effort in environmental control is the recovery systems associated with disassembly and recycling. Efforts by consumer groups and government are encouraging corporations to design and produce environmentally harmless products. This same effort encourages the firms to reengineer their products so when they are to be replaced or discarded, the parts and components can be readily removed for possible reuse in new items. This is taking place particularly in the following type of products computers, printers, copy machines, rifles, pistols, iphones, clocks, watches, lawnmowers, snow blowers, and irons.

Disassembly is also important in the reuser of the mechanisms of electromotive, trucks, buses, automobiles, agriculture, and construction equipment; they include engines, transmissions, drive shafts, clutch, drive axle, driveline, steering gears, and pumps. At some point, these major components of the vehicles are of no more use and are removed and replaced. Instead of scrapping the units, many are purchased by various remanufacturing corporations where they are called cores; and are stored in a warehouse facility. A core is essentially a main component that is capable of remanufacture and worthy for reuse. Note, for an engine, the main core is the engine and the component cores are the components taken out of the engine. When customer demands arise, the cores are taken from the warehouse, disassembled by removing the inside components. The components are cleaned and inspected for future use, repaired if needed, and scrapped if not acceptable. When all the components are available, either from the disassembled unit or from new purchased components, the main core is assembled, tested, and put on sale with a full warranty.

Robotic Assembly A common problem facing assembly line workers is the continual repetitive work that causes physical and mental fatigue. The workers require various breaks during the day for eating, relaxing, and bodily needs. In contrast, robots work for long hours and require no breaks. The robots are capable of performing assembly tasks ranging from routine to precise without losing quality. A robot is a device that is programmed to perform a variety of tasks. With various controls, the assembly engineer steers the robot to perform as needed for the specific task that it is assigned.

Manufacturing robots are costly due to expenses in designing, hardware, operating software, installation, and maintenance. But when compared to hiring and maintaining a human worker, the long-term cost could favor a robotic station. Robots do not require wages, benefits, vacations, insurance, severance pay, pensions, or union demands. Any time after installation, the robot can be upgraded and replaced for an advance model without any worker repercussions. To do the same to a human worker may cause the need for severance benefits or the possibility of a lawsuit.

Quality workmanship improves with the installation of robots in place of humans, as long as the robot is running smoothly and the parts and components of installation are without fault. A problem could occur with poor fixtures or parts that slip out of the grippers of the robot. A maintenance operator is assigned to monitor a series of robot stations in the event of a mishap. In general, the robots are very reliable and require minor maintenance.

Balancing of a robotic assembly line is the process of ensuring the work times at each station will allow the flow of units to pass along smoothly with minimum delays. The plant may have a variety of robot types where each has different capabilities and speeds. The best robot for each station has to be assigned depending on the task of the station and the capabilities of the robots. When a new product is planned for assembly, the assigning of robots to the stations may need to be rearranged. Some of the robots may need to be reprogrammed by guiding its arms and pushing a few buttons.

Robotics is in use in a variety of industries (automotive, aerospace, electromotive, medical, consumer goods, electronics, and industrial). Some assembly lines are partially automated and others are fully automated with robots, and the robots within have applications ranging from handling small to large objects. Common applications are welding, testing, serial part marking, labeling, drilling, cutting, spraying, painting, grinding, molding, material removal, material movement, milling, polishing, bonding, and water jet.

Chapter 3
Some Fundamentals

Introduction

This chapter describes the data, computations and decisions needed to control an assembly line. The key data includes the work elements, the element times, the predecessor elements, the unit time, shift time and the shift schedule quantity. The data is used to compute the number of operators needed on the line, the cycle time, average operator time, balance delay, and the efficiency ratio. The management can assign the work elements to the operators where the operator times are evenly distributed and all precedence constraints are satisfied accordingly. This step is called line balancing. Next, the operator times are measured along with the effective cycle time and effective balance delay. The bill-of-material of the product(s) now enters as data to determine the part and component requirements needed for a shift duration at every station along the line. The chapter also describes the relation between the main line, feeder lines, subassembly lines, and labor groups, as well as the use of parallel stations, and parallel lines.

To build one unit of an item on an assembly line, all of the parts and components listed on the bill-of-material are gathered and are inserted one at a time until the unit of product is complete. This process takes place on an assembly line that includes a series of stations and each with one or more operators. Each operator is assigned a list of tasks in the process. The parts are placed along the line where the associated assembly tasks are performed. The unit moves from one station to the next and when it leaves the last station, the unit is a finished good item.

Work Elements

For convenience here, h designates the part (or component), Nh identifies the number of parts and $h = 1$ to Nh gives the list of the parts on the bill-of-material. The tasks of assembly are defined and are called work elements, or simply elements. Altogether, Ne designates the number of elements and for simplicity here,

N. T. Thomopoulos, *Assembly Line Planning and Control*,
DOI: 10.1007/978-3-319-01399-2_3,

the elements are labeled as $e = 1$ to Ne. Each element identifies the task to be performed and the standard time of the task, t_e. For simplicity in the text, $t = t_e$ is often used to denote the element standard time. Should the task of the elements include the first attachment of a part on the unit, the part, h, is also listed along with the element. The element should also include the immediate predecessor element(s), if any, that need to be completed before the element task can begin. So for each element, the instructions include the following: element number, task, standard time, immediate predecessor element(s), part(s). The element task is the description of the element and is not included in the examples of this test.

Precedence Diagram

It is convenient to draw a chart, called a precedence diagram, to depict the build relationship among all the elements. The diagram shows which elements can begin without any predecessor elements, and which elements have predecessor elements. The precedence diagram is like a blueprint on how to assemble the unit. An example of Ne = 10 elements and any associated predecessor elements is listed in Table 3.1.

The corresponding precedence diagram is shown in Fig. 3.1. The flow is from left to right. The numbers denote the elements and the connecting lines identify the immediate predecessor elements.

Unit Time

Assume the standard work element times, t_e, in Table 3.1 are listed in minutes. The sum of the standard times represents the total time to complete one unit of product and is called the unit time and denoted as Σt_e. In the example, this is $\Sigma t_e = 14.5$ min.

Table 3.1 Ten elements e, standard times t_e, and their predecessor elements

e	t_e	Predecessor elements
1	1.1	2
2	2.6	
3	1.8	2
4	2.0	1
5	2.3	2
6	0.5	4, 5
7	1.4	3, 5
8	0.8	6
9	1.2	8
10	0.8	7, 9

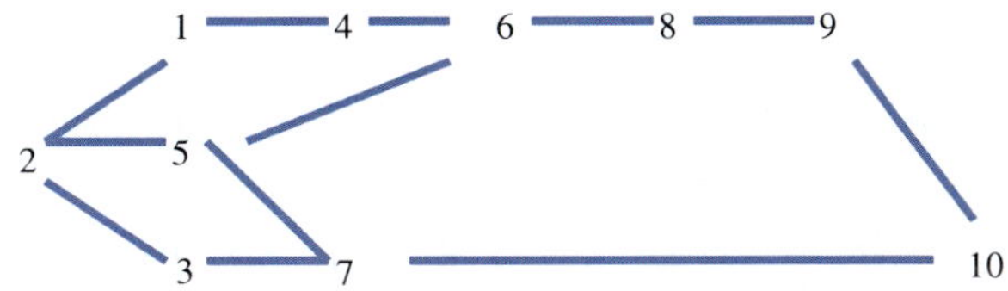

Fig. 3.1 A ten element precedence diagram

Shift Time

The shift time, T, represents the total working time during one shift. Suppose a shift is from 8 a.m. to 4:30 p.m. where the operators takes a 30 minute lunch break at noon and 15 minute breaks in the morning and again in the afternoon. In this situation the shift working time becomes $T = 450$ min.

Shift Schedule

The shift schedule, N, identifies the desired number of units of the finished good item to complete during the shift. For example, should the schedule call for ninety units per shift, then $N = 90$.

Number of Operators

The minimum number of operators, n, needed to accomplish the schedule is computed as follows:

$$n = \sum t_e \times N/T$$

In the example with $\Sigma t_e = 14.5$ min, $N = 90$ units, and $T = 450$ min,

$$n = 14.5 \times 90/450 = 2.9$$

When n is not an integer, the value is rounded up to the closest integer, $n = 3$, in the example.

Cycle Time

The cycle time, denoted as c, is a measure of the time between two units coming off the end of the line as finished good items. This measure is computed as

$$c = T/N$$

where T is the shift time and N is the shift schedule. In the example where $T = 450$ min and $N = 90$ units,

$$c = 450/90 = 5.00\,\text{min}.$$

Average Operator Time

The average time assigned to the operators is denoted as $\bar{c}$ and is called the average operator time. This is computed by

$$\bar{c} = \sum t_e/n$$

where Σt_e is the unit time for one finished good item and n is the number of operators on the line. In the example with $\Sigma t_e = 14.5$ and $n = 3$, the average operator time becomes,

$$\bar{c} = 14.5/3 = 4.83.$$

Balance Delay

A way to measure the efficiency of the line is to compute the portion of idle time per unit. This is called the balance delay and is denoted as d. The balance delay is obtained by the cycle time, c, and the average operator time, $\bar{c}$, as follows,

$$d = (c - \bar{c})/c$$

In the example where $c = 5.00$ min and $\bar{c} = 4.83$ min,

$$d = (5.00 - 4.83)/5.00 = 0.034$$

and so, the percent of idle time is $d = 3.4\ \%$.

Efficiency Ratio

Another common measure is called the efficiency ratio, E, and is computed by

$$E = \bar{c}/c$$

This is a measures of the portion of the time the operators are busy working on the units. In the example, where $c = 5.00$ and $\overline{c} = 4.83$,

$$E = 4.83/5.00 = 0.966$$

or $E = 96.6$ %.

Note the relation between the balance delay, d, and the efficiency ratio, E, is

$$d = 1 - E.$$

In the example,

$$d = 1 - 0.966 = 0.034$$

Line Balance

An important step for the assembly management is to assign the work elements, e, to the n operators on the line. This entails spreading the work time as evenly as possible to the operators in a way where the elements do not violate any of the precedence restrictions. This process is called line balancing.

In the example of Table 3.1 and Fig. 3.1, Ne = 10 elements, $\overline{c} = 4.83$ min, $\Sigma t_e = 14.5$ min and $n = 3$ operators. One assignment of the elements is listed in Table 3.2.

Operator Times

The time assigned to each of the n operators on the line is called the operator times and is denoted as c_i, $i = 1$ to n. Note below where the sum of the operator times equals the unit time, i.e,

Table 3.2 Operator i, work element e, with work element time t_e, and operator i time, c_i

i	e	t_e	c_i
1	2	2.6	
	5	2.3	4.9
2	1	1.1	
	3	1.8	
	4	2.0	4.9
3	6	0.5	
	7	1.4	
	8	0.8	
	9	1.2	
	10	0.8	4.7

$$\sum c_i = \sum t_e$$

In the example, $\Sigma c_i = (4.9 + 4.9 + 4.7) = 14.5$ min, and $\Sigma t_e = 14.5$ min.

Effective Cycle Time

The effective cycle time can now be measured since the station operator times are known. Note where the units cannot be processed down the line any faster than the largest operator time, and because of this, the cycle time becomes the maximum of the operator times. That is,

$$c = \max(c_1, \ldots., c_n).$$

Continuing with the example, and using the line balance results from Table 3.2, the effective cycle time becomes,

$$c = \max(4.9, 4.9, 4.7) = 4.9\,\text{min}.$$

Effective Balance Delay

The effective balance delay can be computed when the effective cycle time is known. In the example where $\overline{c} = 4.83$ and $c = 4.90$, the effective balance delay is the following:

$$d = (4.90 - 4.83)/4.90 = 0.014$$

The corresponding efficiency, E, of the line becomes,

$$E = 4.83/4.90 = 0.986.$$

Bill of Material

The bill-of-material (bom) on a product lists all the parts and components needed on each unit of a product and the quantity of each. Suppose the finished good item of the ten-part example requires five parts (or components). For clarity, the following definition is used here: the first element to use part, h, is called the prime element of the part, denoted as $e(h)$, and the number of pieces of part, h, that is needed on the unit is called the bom quantity per unit, denoted at $b(h)$. The Nh = 5 parts are listed on the bill-of-material in Table 3.3. The tables show the parts, denoted as h, the prime element, $e(h)$, and the bom quantity per unit, $b(h)$. For simplicity in the text, the notation often uses $b = b(h)$ and $e = e(h)$. Hence, part $h = 1$ is first used by element $e = 1$ and one unit of $h = 1$ is needed per unit of product. Part $h = 4$ is first used with element $e = 5$ and requires two units for each unit of product, so forth. For convenience in much of the remainder of the book, the notation for $e(h)$ will be e, and for $b(h)$ it is simply b.

Part Requirements

Continuing with the example of Table 3.1, three stations are assigned work and the line balance results are listed in Table 3.2. With this information, and the bill-of-material of Table 3.3, the shift requirements by part, R_h, for each station is listed in Table 3.4. The table lists the following: station, i, element, e, part, h, and the quantity per unit, $b(h)$. The table shows the shift requirements of part, h, to have available at the station at the start of the shift. Note, $R_h = N = 90$ for every part where $b(h) = 1$, and is $R_h = 2N = 180$ for part $h = 4$ where $b(h) = 2$.

Stations and Operators

The terms stations and operators on the line are often used with the same meaning. In general, the number of operators on a line is greater of equal to the number of stations on a line.

Table 3.3 List of parts h, with prime elements $e(h)$, and bom quantity per unit $b(h)$

h	e(h)	b(h)
1	1	1
2	2	1
3	3	1
4	5	2
5	6	1

Table 3.4 Operator i, element e, part h, bom quantity per unit $b(h)$, and shift requirements, R_h

i	e	h	b(h)	R_h
1	2	2	1	90
	5	4	2	180
2	1	1	1	90
	3	3	1	90
	4			
3	6	5	1	90
	7			
	8			
	9			
	10			

When a line has n operators, the line may also be known as n stations. This occurs frequently when each operator has it's own location on the line, whereby the units passing through are in a station of only one operator. In this situation, the unit is in the possession of only one operator at a time.

In other situations, the station may have two or more operators. This may occur should the workload in the station require more than one operator to carry out the task assigned. On a truck assembly line, it may take two operators to work together to install the cab of the truck; one operator on the left-hand side of the truck, and the other on the right-hand side.

Main Line, Subassembly Lines, and Labor Groups

Some assembly lines have only one beginning and ending location. The assembly of the unit starts at station 1 and ends, as a finished good item, at station n, and there are no subassembly lines.

Another situation is when a line has two or more labor groups, where the work elements performed in one labor group is mutually exclusive to the work elements in another labor group. For example, units going down a line are assembled one-by-one. They eventually finish at the last station, and then enter a paint booth where the units are spray painted, with another set of operators. The operators in the initial line belong to labor group 1, and the operators in the paint booth are in labor group 2. In this situation, the main line consists of two labor groups.

In other situations, one line, called the main line, is used to start the unit and ends when the unit is a finished good item, but requires one or more subassembly lines. The units coming off the subassembly lines are components of the units on the main assembly line. These subassembly lines are often called feeder lines, where components of the finished good item are assembled separately and subsequently joined to the unit on the main assembly line. The workers on the main line and in the subassembly lines are in separate labor groups since the work in each of the labor groups is mutually exclusive from each other.

Figure 3.2 is an example of a truck assembly system with six labor groups. The main line has three labor groups (frames and axles, paint, completion) and three subassembly lines (engines, cabs, tires). Note, the subassembly lines are also separate labor groups.

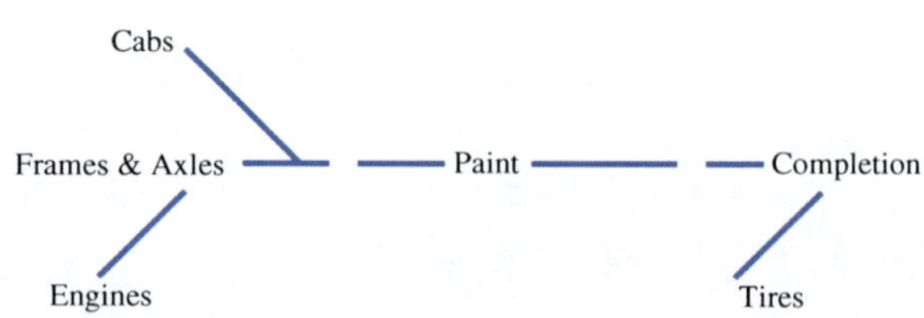

Fig. 3.2 A main line with six labor groups and three subassembly lines

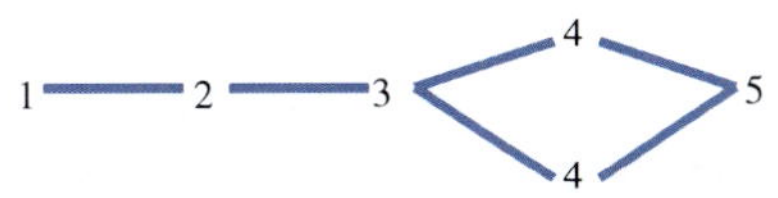

Fig. 3.3 The flow of units down the line with station 4 in parallel

Parallel Stations

In some lines, one or more of the work elements have a large work time and such an element(s) disrupts the flow of the entire line. For example, the management would be confronted in what to do should one element standard time be 8.1 min on a line when the desired cycle time is half that size? This situation calls for the use of a parallel station. Two identical stations are setup next to each other, and as the units flow down the line, every other unit will stop in one of the stations. This way, the units will flow out of (one of the stations) each 8.1/2 = 4.05 min.

Consider the example when five stations with operator times: (4.1, 4.0, 3.9, 8.1, 3.7) min. The units in station 4 require 8.1 min and the time for the units in the other stations are about half that. The solution is to have two parallel work places at station 4, where every other unit coming out of station 3 will go to one of the two locations of station 4. This way, the cycle time for the line becomes the following:

$$c = \max(4.1, 4.0, 3.9, 8.1/2, 3.7) = 4.1 \text{ min.}$$

See Fig. 3.3 for the flow of units down the line.

Continuing with the example, the number of stations is five—two are parallel—the number of operators is $n = 6$, the cycle time is $c = 4.1$ min, the unit time is $\Sigma t_e = 23.8$ min, the average operator time is $\overline{c} = 3.97$ min and the balance delay is $d = 0.032$.

Summary

This chapter describes some of the basic fundaments of the assembly operation in a plant. The discussion begins with the tasks needed to produce one unit of product on the assembly line. The tasks are called work elements, and each element has a standard time and most have one or more predecessor elements. The collection of elements is depicted in a precedence diagram. The shift time and the shift schedule quantity are needed along with the sum of the work element times to determine the number of operators to have on the assembly line. Next, the elements are assigned to the operators in a way where the operator times are evenly distributed, and all precedence constraints adhered. This process is called line balancing. Measures are described that allow the management to determine the efficiency of the line. The bill-of-material specifies the parts needed in the assembly of one unit of product, and this information is used to determine the requirements of each part over a shift's duration. The part requirements for the shift are stocked at each operator station to enable the worker to carryon his/her tasks on the product. When one or more elements have an excess of element time, parallel stations may be used on the line. Oftentimes the assembly process has a main line that is connected with subassembly lines and feeder lines.

Chapter 4
Assembly Planning

Introduction

Through examples of single model and mixed model assembly lines, this chapter shows how to make the best decisions in order to meet the production plans specified while attaining efficiency in the assembly operation for future time periods. An example of a single model line shows the data and computations that allow management to select the production schedule for the line. A range of shift schedule quantities are considered and with each quantity, the number of operators needed on the line and the associated efficiency measures are calculated. The first example is a single model assembly situation with one labor group, and the next contains two labor groups. The computations give the speed of the conveyor system and also the length requirements of the assembly line. When more than one labor group is needed, the cycle times and conveyor speeds need to be coordinated, so the units flow appropriately from one labor group to the next. Examples of mixed model make-to-stock assembly with one labor group and with five labor groups demonstrate the data and computations required to determine the number of operators to assign in each labor group, and how to measure the operation efficiencies.

Some assembly lines are flexible where a variety of products are assembled on the same facility. Say, product A is set up on the line for a period of time, several days to several weeks. When the inventory on A is built up to satisfy the foreseeable demands, the line changes over to accommodate product B. In a later day, product A will again be set up on the line. In this way, this is a batch assembly line that accommodates the many products. Although within each batch, the line is essentially a single model line, handling one product at a time.

Single Model Lines

The assembly management must do some preliminary planning for the line. How many units, N, to schedule for the shift time, T, is an important consideration. The combination of shift time, shift schedule, and the product unit time, Σt_e, are

N. T. Thomopoulos, *Assembly Line Planning and Control*,
DOI: 10.1007/978-3-319-01399-2_4,

ingredients that allow the management to determine the number of operators needed on the line. This then leads to the cycle time. The space needed for one unit of product is here called the unit space and denoted as *w*. This is the distance between equal points in two units on the line. For example, this is the distance from a fixed point of one unit to the same fixed point of the next unit on the line. This helps determine the conveyor speed, and subsequently the length of the line. In the event the units move down the line steadily in a conveyor type device, the conveyor speed is determined. Also, the length of the line that is needed can be measured. The management wants to set the shift schedule and the number of operators in a way where the line runs in an efficient way and has minimal idle time, measured by the balance delay, *d*.

Labor Groups

When the line has one or more feeder lines, more complications take place. One line is like a sub assembly that produces a component that is attached to the main unit. There could be a need for one unit of the component from the feeder line, to one unit of the main item on the main line. Or there may be a need for two or more such components for each unit on the main line. The shift schedule of each feeder line must be synchronized with the main line so that the output flow of units down all the lines are in compliance. The feeder line has its own shift schedule, number of operators, cycle time, unit space, conveyor speed and line length, and each line has its own labor group. Each labor group has operators of its own and they do not move from one line to another. The notation for the labor groups is *g*.

Mixed Model Lines

The assembly management has even more decisions to make in setting up a line for mixed model assembly. In a make-to-stock line, *Nj* models are produced on the line and over a shift time, *T*, the number of units to schedule on the line is needed to meet the oncoming demands from the customers. Each model on the line has different unit times and altogether the mix of the model schedule often varies over time. The management must determine the shift schedule to meet all the needs of the models in a way that allows the line to run smoothly and efficiently. Line planning determines the shift schedule, the number of operators needed, the conveyor speed and the line length.

When feeder lines are included in the mixed model assembly, more considerations are needed. The schedules for each of the feeder lines (labor groups) are determined, so the flow of the entire system is synchronized to meet the needs of the main line. The shift schedule by labor group, the number of operators, the conveyor speed and the line length are all needed for each line in the system.

This chapter describes four situations where line planning is used. These are the following:

Single model assembly with one labor group.
Single model assembly with five labor groups.
Mixed model make-to-stock assembly.
Mixed model make-to-stock assembly with two labor groups.

Single Model Assembly with One Labor Group

Consider a single model line with n stations and one operator per station. The units go straight down the line and the assembly begins at station 1 and ends at station n, and thereby there is only one labor group. Assume that the shift time is $T = 450$ min, the unit time is $\Sigma t_e = 50$ min, and shift production schedules from $N = 90$ to 110 units are under review. Suppose further, the length of a unit is one foot and the average length of a station is 4.00 ft. The example assumes that the length separating an equal point on the units is the same as the station length of $w = 4.00$ ft, also called the unit space. The management is seeking computations that show the feasibility on running the line from 90 to 110 units per shift.

Table 4.1 is a planning guide for this line. Results are listed for each shift schedule from $N = 90$ to 110 units. The column notations are the following:

N = shift schedule quantity
n' = raw number of operators needed
n = integer number of operators needed
c' = cycle time
$\bar{c}$ = average operator time
d = balance delay
v = conveyor speed
L = line length

For example, when the shift schedule is 100 units, the row with $N = 100$ lists the following:

$n' = N \times \Sigma t_e/T = 100 \times 50/450 = 11.11$ operators
$n = 12$ operators
$c' = T/N = 450/100 = 4.50$ min
$\bar{c} = \Sigma t_e/n = 50/12 = 4.17$ min
$d = (c' - \bar{c})/c' = (4.50 - 4.17)/4.50 = 0.073 = 7.3\,\%$
$v = w/c' = 4.00/4.50 = 0.89$ ft per minute
$L = w \times n = 4.00 \times 12 = 48$ ft

Note, the most (potentially) efficient situations occur with the minimum measures of balance delay, d. In the table, the minimum is $d = 0.00$ and this occurs when the shift schedule is $N = 90$, 99, and 108 units. Caution, however, because

Table 4.1 Planning guide for a single model line with one labor group $T = 450$ min, $\Sigma t_e = 50$ min, $w = 4$ ft, $N = 90$–110

N	n'	n	c'	$\bar{c}$	d	v	L
90	10.00	10	5.00	5.00	0.00	0.80	40.00
91	10.11	11	4.95	4.55	0.08	0.81	44.00
92	10.22	11	4.89	4.55	0.07	0.82	44.00
93	10.33	11	4.84	4.55	0.06	0.83	44.00
94	10.44	11	4.79	4.55	0.05	0.84	44.00
95	10.56	11	4.74	4.55	0.04	0.84	44.00
96	10.67	11	4.69	4.55	0.03	0.85	44.00
97	10.78	11	4.64	4.55	0.02	0.86	44.00
98	10.89	11	4.59	4.55	0.01	0.87	44.00
99	11.00	11	4.55	4.55	0.00	0.88	44.00
100	11.11	12	4.50	4.17	0.07	0.89	48.00
101	11.22	12	4.46	4.17	0.06	0.90	48.00
102	11.33	12	4.41	4.17	0.06	0.91	48.00
103	11.44	12	4.37	4.17	0.05	0.92	48.00
104	11.56	12	4.33	4.17	0.04	0.92	48.00
105	11.67	12	4.29	4.17	0.03	0.93	48.00
106	11.78	12	4.25	4.17	0.02	0.94	48.00
107	11.89	12	4.21	4.17	0.01	0.95	48.00
108	12.00	12	4.17	4.17	0.00	0.96	48.00
109	12.11	13	4.13	3.85	0.07	0.97	52.00
110	12.22	13	4.09	3.85	0.06	0.98	52.00

the effective balance delay depends on how well the line can actually be balanced, where the operators are assigned their work elements on the units. When perfectly balanced, the effective cycle time, c, would be the same as c'. Otherwise $c \geq c'$ and thereby the effective balance delay could be larger than the d values listed on the table.

Note further, the conveyor speed is computed by $v = w/c'$. This measure only applies when the units are continuously moving from one station to the next along the line, like on a conveyor belt. In the event, the operators push their completed units to the next station, the conveyor speed, v, does not apply.

The table also lists the length of the line, L, with each shift schedule size. These measures range from 40 to 52 ft, assuming the line is in one continuous stretch from station 1 to station n.

With the table results listed, the management can plan accordingly on how many units to schedule for a shift. The schedule size depends mostly on the requirements for the units from the downstream locations of warehouses, distribution centers, retailers, and so forth. Sometimes, extra units are planned in the assembly to account for possible defectives or units for service part needs.

Single Model Assembly with Five Labor Groups

Now assume a system with a main line and four feeder lines, called labor groups. These are five separate lines whose workers are dedicated to their assigned group and are not shared between groups. Hence, there are five labor groups. See Fig. 4.1.

Assume the shift time of $T = 450$ min is the same for all five labor groups ($g = 1, 2, 3, 4, 5$). The shift schedule, N, is based on the main line ($g = 1$), where management calls for a range of $N = 97$–103. The data available for each of the lines are listed in Table 4.2.

The assembly planning for this system is summarized in Table 4.3, where the results cover the five labor groups over the seven options of shift schedules ($N = 97$ to 103) for $g = 1$. The entries in the table are the following:

g = labor group
b = bill-of-material units
N = shift schedule units
Σt_e = unit time (minutes)
n' = raw number operators needed
n = integer number operators needed
c' = cycle time (minutes)
$\overline{c}$ = average operator time (minutes)
d = balance delay
w = unit space (feet)
v = conveyor speed (feet/minute)

Note in the table when $N = 97$ at $g = 1$, the shift schedule for each labor group is $b \times 97$, where b is the bom number of units needed in each group g for every unit in $g = 1$. For example, at $g = 2$, $b = 2$ and thereby $N = 2 \times 97 = 194$ units. In general, $N_2 = b_2 \times N_1$. Some other computations at $g = 1$ are the following:

$n' = N\Sigma t_e/T = 97 \times 50/450 = 10.78$ operators
$n = 11$ operators
$c' = T/N = 450/97 = 4.64$ min
$\overline{c} = \Sigma t_e/n = 50/11 = 4.55$ min
$d = (c' - \overline{c})/c' = (4.64 - 4.55)/4.64 = 0.02 - 2.0\,\%$
$v = w/c' = 10.00/4.64 = 2.16$ ft per minute

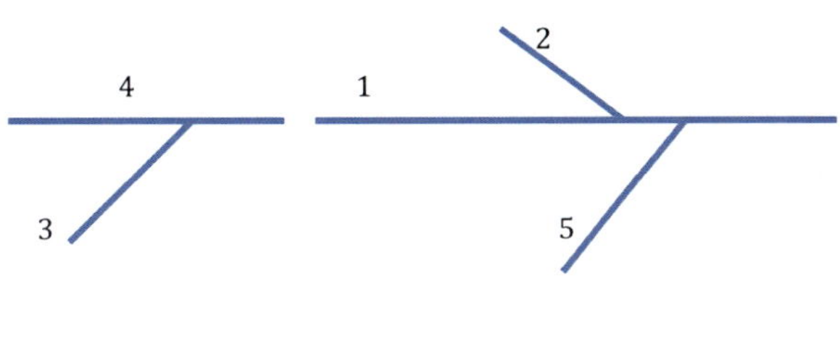

Fig. 4.1 Assembly system with the main line (*1*) and four feeder lines (*2*, *3*, *4*, *5*)

Table 4.2 Labor groups (g) with bill-of-material quantity (b), unit time (Σt_e minutes), and unit space (w feet)

g	b	Σt_e	w
1	1	50	10
2	2	20	5
3	1	25	8
4	1	30	10
5	4	10	5

$L = w \times n = 10.00 \times 11 = 110$ ft

The balance delay for the total system, of five labor groups, is also computed. This is by the relation below where d_g is the balance delay at labor group g, and n_g is the number of operators needed at g.

$$d = \sum_g (n_g d_g) / \sum_g n_g.$$

Note at the first option, where $N = 97$ for $g = 1$, the balance delay for the total system is $d = 0.05$.

The total balance delay over all five labor groups is listed in Table 4.3, and the summary is shown in Table 4.4.

When $N_1 = 97$ for the main line ($g = 1$), the number of operators needed on the total line is $\Sigma n_g = 42$, and the balance delay is $d = 0.05$. The minimum balance delay—in two decimals—for the total system occurs when $N_1 = 99$ and 101 where $d = 0.03$.

The computations assume all five labor groups are on separate conveyor systems. The five labor groups have to be synchronized where the units flowing out of a labor group meets its downstream labor group, if any, as needed. For example, the units flowing out of $g = 4$ are timed as input to $g = 1$. In the same way, the output of $g = 3$ flows to $g = 4$, and the outputs from $g = 2$ and 5 flow, as needed, to $g = 1$. To accomplish, the conveyor speeds are set accordingly. The conveyor speeds are computed using the unit space (w) for each unit on every labor group, g, and the associated cycle time, c'.

With the table results listed, the management can plan accordingly on how many units to schedule for a shift. The schedule size depends mostly on the requirements of the units from the downstream locations (warehouses, distribution centers, and retailers). Sometimes, extra units are planned in the assembly to account for possible defective units.

Mixed Model Make-to-Stock Assembly

Consider now a mixed model line where three make-to-stock models are assembled. The models are denoted as $j = 1, 2, 3$ and the shift schedule N is the number of units to produce for the shift and is the sum of the shift schedules for the three models. Each model j has different unit times, Σt_{ej}, and shift schedule requirements,

Table 4.3 Assembly planning for the single model line with five labor groups $T = 450$ min

g	b	N	Σt_e	n'	n	c'	$\overline{c}$	d	w	v
1	1	97	50	10.78	11	4.64	4.55	0.02	10	2.16
2	2	194	20	8.62	9	2.32	2.22	0.04	5	2.16
3	1	97	25	5.39	6	4.64	4.17	0.10	8	1.72
4	1	97	30	6.47	7	4.64	4.29	0.08	10	2.16
5	4	388	10	8.62	9	1.16	1.11	0.04	5	4.31
								0.05		
1	1	98	50	10.89	11	4.59	4.55	0.01	10	2.18
2	2	196	20	8.71	9	2.30	2.22	0.03	5	2.18
3	1	98	25	5.44	6	4.59	4.17	0.09	8	1.74
4	1	98	30	6.53	7	4.59	4.29	0.07	10	2.18
5	4	392	10	8.71	9	1.15	1.11	0.03	5	4.36
								0.04		
1	1	99	50	11.00	11	4.55	4.55	0.00	10	2.20
2	2	198	20	8.80	9	2.27	2.22	0.02	5	2.20
3	1	99	25	5.50	6	4.55	4.17	0.08	8	1.76
4	1	99	30	6.60	7	4.55	4.29	0.06	10	2.20
5	4	396	10	8.80	9	1.14	1.11	0.02	5	4.40
								0.03		
1	1	100	50	11.11	12	4.50	4.17	0.07	10	2.22
2	2	200	20	8.89	9	2.25	2.22	0.01	5	2.22
3	1	100	25	5.56	6	4.50	4.17	0.07	8	1.78
4	1	100	30	6.67	7	4.50	4.29	0.05	10	2.22
5	4	400	10	8.89	9	1.13	1.11	0.01	5	4.44
								0.04		
1	1	101	50	11.22	12	4.46	4.17	0.06	10	2.24
2	2	202	20	8.98	9	2.23	2.22	0.00	5	2.24
3	1	101	25	5.61	6	4.46	4.17	0.06	8	1.80
4	1	101	30	6.73	7	4.46	4.29	0.04	10	2.24
5	4	404	10	8.98	9	1.11	1.11	0.00	5	4.49
								0.03		
1	1	102	50	11.33	12	4.41	4.17	0.06	10	2.27
2	2	204	20	9.07	10	2.21	2.00	0.09	5	2.27
3	1	102	25	5.67	6	4.41	4.17	0.06	8	1.81
4	1	102	30	6.80	7	4.41	4.29	0.03	10	2.27
5	4	408	10	9.07	10	1.10	1.00	0.09	5	4.53
								0.07		
1	1	103	50	11.44	12	4.37	4.17	0.05	10	2.29
2	2	206	20	9.16	10	2.18	2.00	0.08	5	2.29
3	1	103	25	5.72	6	4.37	4.17	0.05	8	1.83
4	1	103	30	6.87	7	4.37	4.29	0.02	10	2.29
5	4	412	10	9.16	10	1.09	1.00	0.08	5	4.58
								0.06		

Table 4.4 Summary of some total measures, Σn_g (number of operators) and d (balance delay) when N_1 = shift schedule at $g = 1$

N_1	Σn_g	d
97	42	0.05
98	42	0.04
99	42	0.03
100	43	0.04
101	43	0.03
102	45	0.07
103	45	0.06

N_j. The units times, Σt_{ej}, are (50, 55, 60) minutes for models 1, 2, and 3, respectively. The daily shift schedules can vary and on a typical day, model 1 has the most requirements, followed by models 2 and then 3. In a typical situation, a 1-month planning period is in review and the management wants to set the daily number of units to produce at a fixed quantity. Assume the management is considering schedule quantities of 97–101 units. Although the shift schedule will be fixed for the planning period, each day, the mix of the models may vary slightly, depending on the requirements from the downstream locations. Should 100 units be scheduled for a shift, assume the typical model mix would be (61, 30, 9) units for models 1, 2, and 3, respectively.

Table 4.5 is a listing on some of the preliminary computations for the five options of ($N = 97$, 98, 99, 100, and 101) units for the shift schedule. The table contains the following for each option:

j = model
N_j = shift schedule for model j
Σt_{ej} = unit time for model j
$N_j\Sigma t_{ej}$ = shift time needed for model j

The sum of the shift time over the three models is also listed.

Table 4.6 summarizes the assembly planning computations for the five options on shift schedule, N. The table lists the following:

N = shift schedule for all models
$\Sigma_j N_j \Sigma t_{ej}$ = shift time for all models
n' = raw number of operators needed
n = integer number of operators needed
c' = cycle time
$\overline{c}$ = average operation time per operator
d = balance delay
w = unit space
v = conveyor speed

When $N = 97$, the computations are as below:

$$n' = \Sigma N_j \Sigma t_{ej}/T = 5085/450 = 11.30 \text{ operators}$$

Table 4.5 Shift times for a mixed model make-to-stock line of 3 models

j	N_j	Σt_{ej}	$N_j \Sigma t_{ej}$
1	59	50	2950
2	29	55	1595
3	9	60	540
Sum	97		5085
1	60	50	3000
2	29	55	1595
3	9	60	540
Sum	98		5135
1	60	50	3000
2	30	55	1650
3	9	60	540
Sum	99		5190
1	61	50	3050
2	30	55	1650
3	9	60	540
Sum	100		5240
1	61	50	3050
2	30	55	1650
3	10	60	600
Sum	101		5300

Table 4.6 Assembly planning for the shift schedule options, N, for a mixed model make-to-stock line

N	$\Sigma_j N_j \Sigma t_{ej}$	n'	n	c'	$\bar{c}$	d	w	v
97	5085	11.30	12	4.64	4.37	0.06	10	2.16
98	5135	11.41	12	4.59	4.37	0.05	10	2.18
99	5190	11.53	12	4.55	4.37	0.04	10	2.20
100	5240	11.64	12	4.50	4.37	0.03	10	2.22
101	5300	11.78	12	4.46	4.37	0.02	10	2.24

$n = \text{ceiling of } (11.30) = 12$ operators
$c' = T/N = 450/97 = 4.64$ min
$\bar{c} = \Sigma N_j \Sigma t_{ej}/(N \times n) = 5085/(97 \times 12) = 4.37$ min
$d = (c' - \bar{c})/c' = (4.64 - 4.37)/4.64 = 0.058$
$v = w/c' = 10/4.64 = 2.155$ ft per minute

The table results show that $n = 12$ operators are needed on all of the five options. The option with the minimum balance delay ($d = 0.02$) is when $N = 101$ units.

Mixed Model Make-to-Stock Assembly with 2 Labor Groups

Consider again the mixed model make-to-stock line when only one labor group ($g = 1$) where the shift schedule under review is from $N_1 = 97$ to 101 units. The shift times for the units are listed in Table 4.5 and the assembly planning computations are in Table 4.6.

Assume now a second labor group, ($g = 2$) is a feeder line to the main line ($g = 1$) and two units ($b = 2$) coming off of $g = 2$ are needed for each unit of $g = 1$. So, in essence, the schedule range for $g = 2$ is from 194 to 202 units, or, $N_2 = 2 \times N_1$.

The shift times for $g = 2$ are computed in Table 4.7 where the three models ($j = 1, 2, 3$) are listed. Note the unit times for the three models of labor group $g = 2$ are 6.5, 7.0 and 7.5 min, respectively. When, for example, the schedule is $N_1 = 97$ in Table 4.5, the model schedules are (59, 29, 9) for $j = 1, 2, 3$, respectively. In a corresponding way, the shift schedule at $g = 2$ is $N_2 = 2 \times N_1 = 2 \times 97 = 194$. Hence, the model schedules are (118, 58, 18) for the three models. At $N_2 = 194$, Table 4.7 shows the shift time is $\Sigma N_j \Sigma_e t_{ej} = 1308$ min. In the same way, the shift times for all five schedules of $g = 2$ are listed in Table 4.7.

The assembly planning computations for the total assembly system of labor groups $g = 1$ and 2 are shown in Table 4.8. Note, the shift schedules for $g = 1$ range for $N_1 = 97$ to 101, and the corresponding shift schedules for $g = 2$ are $N_2 = 2 \times N_1$.

At the first option ($N_1 = 97$), the number or operators needed for $g = 1$ and 2 are 12 and 3, respectively, or 15 altogether. The balance delays are $d = 0.06$ and 0.03 for $g = 1$ and 2, and is $d = 0.05$ for the total system. The unit spaces are 10 and 4 ft for $g = 1$ and 2, and the associated conveyor speeds become 2.16 and 1.72 ft per minute, for $g = 1$ an 2, respectively. In the same way, the table gives the comparative results when $N = 98$, 99, 100, and 101.

Table 4.8 shows the minimum balance delay for the assembly system is when $N_1 = 100$ units whereby $d = 0.02$. The number of operators needed for this option is $n = 15$. The table serves as a planning guide to the management as they decide which schedule option to select.

Summary

This chapter describes the data and computations that allow the management to determine the resources to satisfy the production needs and do this with efficiency in the assembly operation. The methods apply for both single model and mixed model assembly lines. The data are the shift time, the shift schedule quantity and the standard time to complete one unit of product. A range of shift schedule

Table 4.7 Shift times of labor group 2 for model j, shift schedule N_j, unit time $\Sigma_e t_{ej}$ and shift time $N_j \Sigma_e t_{ej}$

j	N_j	$\Sigma_e t_{ej}$	$N_j \Sigma_e t_{ej}$
1	118	6.50	767
2	58	7.00	406
3	18	7.50	135
Sum	194		1308
1	120	6.50	780
2	58	7.00	406
3	18	7.50	135
Sum	196		1321
	120	6.50	780
2	60	7.00	420
3	18	7.50	135
Sum	198		1335
1	122	6.50	793
2	60	7.00	420
3	18	7.50	135
Sum	200		1348
1	122	6.50	793
2	60	7.00	420
3	20	7.50	150
Sum	202		1363

Table 4.8 Planning guide for a mixed model, make-to-stock assembly with 3 models (j) and 2 labor groups (g)

g	N_g	$\Sigma_j N_j \Sigma t_{ej}$	n'	n	c'	$\bar{c}$	d	w	v
1	97	5085	11.30	12	4.64	4.37	0.06	10	2.16
2	194	1308	2.91	3	2.32	2.24	0.03	4	1.72
All				15			0.05		
1	98	5135	11.41	12	4.59	4.37	0.05	10	2.18
2	196	1321	2.94	3	2.30	2.24	0.02	4	1.74
All				15			0.04		
1	99	5190	11.53	12	4.55	4.37	0.04	10	2.20
2	198	1335	2.97	3	2.27	2.24	0.01	4	1.76
All				15			0.03		
1	100	5240	11.64	12	4.50	4.37	0.03	10	2.22
2	200	1348	3.00	3	2.25	2.24	0.00	4	1.78
All				15			0.02		
1	101	5300	11.78	12	4.46	4.37	0.02	10	2.24
2	202	1363	3.03	4	2.23	1.68	0.24	4	1.80
All				16			0.07		

quantities are considered and for each quantity a series of computations determine the number of operators needed, the cycle time, conveyor speed, the line length and efficiency measures of the line.

Chapter 5
Inventory Requirements

Introduction

An important step in planning the assembly of a product is ensuring all of the parts on the bill-of-material are available at the station locations at the start of the assembly process. This chapter shows how to determine the part requirements for single model lines, for mixed model make-to-stock lines, and for the mixed model make-to-order lines. Sometimes the station storage space is limited and therefore just the right quantity of stock is necessary. All of the inventory can be stocked at each station location at the start of a shift, or can be stocked two or more times during the shift in a just-in-time manner. In either event, the requirements are stocked prior to their need on the line. The chapter also describes how to control the daily part replenishments coming from the suppliers. When the assembly line for a product is run day after day, the part requirements over the planning horizon are computed for forthcoming shifts. The future part replenishments use the projected daily shift requirements to determine when, and how much, new replenish stock is needed. The replenish projections over the planning horizon for each part is useful information to the part suppliers allowing them to plan their production activities accordingly.

Single Model Assembly

Each finished good item has a bill-of-material (bom) that defines the parts and components needed to create one unit of product. The assembly of the unit is the process of attaching all the parts and components together in the proper manner as designed by the product engineers. In a single model assembly line, only one product is assembled on the line, one after the other. The bom is known and the parts needed to carryout the shift schedule are gathered before the assembly begins.

N. T. Thomopoulos, *Assembly Line Planning and Control*,
DOI: 10.1007/978-3-319-01399-2_5,

For convenience, the parts on the bom are denoted as $h = 1, 2, \ldots, Nh$, where Nh identifies the number of parts on the bom. The bill-of-material lists all the parts, $h = 1, \ldots, Nh$, and also the quantity of parts, b_h needed on one unit of product. If part $h = 1$ has $b_1 = 1$, one unit of $h = 1$ is needed on the unit of product; and if another part, say $h = 2$, requires four unit on the product, $b_2 = 4$, and so forth. Note, b_h is sometimes denoted as, $b(h)$ and sometimes merely as b.

When the shift schedule for the single model line calls for N units, the inventory requirements on each part over the duration of the shift are computed as follows:

$$\mathrm{R}_h = \mathrm{b}_h \times N \quad h = 1 \text{ to Nh}$$

Example 5.1 Suppose a product X has a bom of ten parts, $Nh = 10$, as listed in Table 5.1. Note where five units of $h = 2$ are needed for each unit of product X. Also two units of $h = 5$, and four units of $h = 8$ are used on product X. One unit of all other parts is needed on each unit of X.

Assume the shift schedule calls for $N = 100$ units of product X for a shift duration. To comply with the shift needs, the management must arrange for the part requirement needs to accommodate the schedule needs. Table 5.2 is a list of the inventory needed for the shift.

Mixed Model Make-to-Stock Assembly

Now consider a mixed model make-to-stock assembly line system. The models are denoted as $j = 1, \ldots, Nj$ where Nj identifies the number of models. Each model has a bill-of-material and many parts are common for the models and some are unique to one or more models. The notation, $u_{hj} = (0,1)$ is a usage index that is associated with each part h and model j. This index is set to one when model j uses part h, and is zero when not. To compute the part requirements for the shift, R_h, the following data is gathered. The list of parts, h, and the bom quantity of each that are needed on a unit of product b_h, is listed for each of the parts $h = 1$ to Nh.

Table 5.1 Bill-of-material for product X

h	b_h
1	1
2	5
3	1
4	1
5	2
6	1
7	1
8	4
9	1
10	1

Table 5.2 Inventory requirements R_h, of part h, for the shift schedule of product X at $N = 100$

h	b_h	R_h
1	1	100
2	5	500
3	1	100
4	1	100
5	2	200
6	1	100
7	1	100
8	4	400
9	1	100
10	1	100

The shift schedule for each of the models, $N_j, j = 1$ to Nj, are also gathered. The part h requirements for the shift are computed as follows:

$$R_h = \sum_j \left[N_j \times u_{hj} \times b_h\right]$$

Example 5.2 Suppose the ten parts of Table 5.1 again with bom data on the parts needed for a make-to-stock, mixed model assembly line with three models, $j = 1$, 2, 3. Assume the usage indices, u_{hj}, for each of the parts are those listed in Table 5.3.

Continuing with the example, assume the shift schedules for each of the models are the following: $N_1 = 50$, $N_2 = 30$ and $N_3 = 20$. With this information, the part requirements for the shift can now be computed as in Table 5.4. Note, for part $h = 1$, $R_1 = (50 + 0 + 20) \times 1 = 70$ pieces. Part 2 requires $R_2 = (50 + 30 + 20)5 = 500$ pieces, and so forth.

Table 5.3 Usage index u_{hj}, of part h, and model j

	——Model j——		
h	1	2	3
1	1	0	1
2	1	1	1
3	1	1	0
4	1	1	1
5	0	1	1
6	1	1	1
7	1	0	1
8	0	1	0
9	0	0	1
10	1	1	1

Note for example, part $h = 1$ is only used on models $j = 1$ and 3

Table 5.4 Inventory requirements, R_h, for each part, h,over the shift schedule, with bom quantity per part, b_h, and model j usage index, u_{hj}

			——Model j——			
		N_j	50	30	20	
h	b_h	u_{hj}	1	2	3	R_h
1	1		1	0	1	70
2	5		1	1	1	500
3	1		1	1	0	80
4	1		1	1	1	100
5	2		0	1	1	100
6	1		1	1	1	100
7	1		1	0	1	70
8	4		0	1	0	120
9	1		0	0	1	20
10	1		1	1	1	100

Mixed Model Make-to-Order Assembly

Now consider a mixed model make-to-order assembly line system. The units to assemble are called jobs and are denoted as j (not to be confused with models). For a shift schedule, N jobs are assigned for assembly. Each job has a set of features, f, and options, k. In essence, every job has a unique bom. For a particular shift, the frequency of options and features are tallied and the sum by f and k is denoted as n_{fk}.

When Same Part for all Options of Feature f

Each job has a bill-of-material and many parts are common for every job and some are not, depending on the combination of features and options. Some of the parts are associated with a particular feature f and option k. Others are not connected with the features. Those parts not associated with features are needed on every job in the schedule. Those parts aligned with a feature and option, are needed on only those jobs with the same f and k combination. Note, $k = 0$ for a feature is called when the job does not use the feature on the unit, and hence, no part is needed on the job. Altogether, N jobs are assigned for the shift and n_{fk} is the number of jobs with the combination of feature f and option k.

The part h requirements for the shift are computed as follows:

$R_h = N \times b_h$ for part h without a feature

$R_h = \sum_{k \geq 1} [n_{fk}] \times b_h$ for part h with feature f

Table 5.5 Option frequency, n_{fk}, for feature f and option k

	n_{fk}				
f \ k	0	1	2	3	4
1	20	80			
2	0	30	20	40	10
3	50	40	10		

Table 5.6 Inventory requirements R_h, for each part h, over the shift schedule when $N = 100$

			Option frequency (n_{fk})					
h	b_h	f/k	0	1	2	3	4	R_h
1	1							100
2	5							500
3	1	1	20	80				80
4	1							100
5	2	2	0	30	20	40	10	200
6	1							100
7	1							100
8	4							400
9	1	3	50	40	10			50
10	1							100

Example 5.3 Suppose the ten parts of Table 5.1 again with bom data. Assume the option frequencies, n_{fk}, over the shift duration are those listed in Table 5.5 where the shift schedule is $N = 100$.

Note, at feature $f = 1$, 20 jobs do not use the feature and 80 do. For feature $f = 2$, 30 jobs call for option 1, 20 for option 2, 40 for option 3, and 10 for option 4. Finally, for feature $f = 3$, 50 jobs do not use the option, 40 call for option 1, and 10 for option 2. The computations for the inventory requirements are shown in Table 5.6.

When Different Part for Each Option of Feature f

Suppose the situation where each option k of a feature f calls for a different part. For notation sake, when part h is associated with a feature, and the option is $k \geq 1$,the part h is denoted as $h.k$. As before, N jobs are assigned for the shift and n_{fk} are the number of jobs with the combination of feature f and option k.

The part h requirements for the shift are computed as follows:

$R_h = N \times b_h$ for part h without a feature
$R_{h.k} = n_{fk} \times b_h$ for part $h.k$ with feature f and option $k \geq 1$
$R_{h.0} = 0$ for part h with feature f and option $k = 0$

Table 5.7 Inventory requirements R_h, for each part h over the shift schedule when $N = 100$, b_h = bom quantity, f, k = feature and option, n_{fk} = number of f and k

h	b_h	f, k	n_{fk}	R_h
1	1			100
2	5			500
3.0	1	3,0	20	0
3.1	1	3,1	80	80
4	1			100
5.0	2	5,0	0	0
5.1	2	5,1	30	60
5.2	2	5,2	20	40
5.3	2	5,3	40	80
5.4	2	5,4	10	20
6	1			100
7	1			100
8	4			400
9.0	1	9,0	50	0
9.1	1	9,1	40	40
9.2	1	9,2	10	10
10	1			100

Example 5.4 Suppose the ten parts of Table 5.1 again are the parts needed for a make-to-order, mixed model assembly line with $N = 100$ jobs. In this situation, each part with a feature f and option $k \geq 1$ is labeled as $h.k$. Assume the option frequencies, n_{fk}, over the shift duration are those listed in Table 5.5.

The computations for the inventory requirements are the same as shown in Table 5.7.

Inventory Replenishments

Having computed the inventory requirements for each part used on the assembly line, the next step is to determine the replenishment schedule for the parts. When the inventory on a part is high, no replenishment is needed, and when not high, a replenishment quantity is needed. This step has to be taken daily to ensure the supply of parts is available to complete the assembly needs of the lines. The discussion below shows a way to control the inventory for the parts.

Part Data

To determine the replenishment quantity for a part, the data needed are the following:

h = part
R_h = part requirement for the shift duration.
R_{ht} = part h requirement for future days t over the planning horizon.
In the analysis here, assume $R_{ht} = R_h$.
OH = on-hand inventory
OO = on-order inventory
M = multiple quantity from supplier
L = lead time from supplier in days
P_{ss} = parameter of safety stock in days
P_{buy} = parameter of buy quantity in days.

When the assembly schedules are known for more than one day, the requirements are computed as described earlier in this chapter. But, when not known, the requirements for the days $t > 1$ are assumed the same as when $t = 1$. The inventory on-hand is denoted as OH and the inventory on-order is OO. Should the supplier only ship the part in a multiple quantity M, the buy quantity must be in multiples of M. The supplier specifies the lead time and this is labeled as L (days). The assembly management furnishes two parameters for all the parts. These are P_{ss} = days of safety stock to have available, and P_{buy} = days of the buy quantity.

Order Point and Order Level

With the above part data, it is now possible to compute the replenish measures, order point, OP, and order level, OL. These are computed in the following way.

$\text{OP} = (L + P_{ss})$ future days supply
$\text{OL} = \left(L + P_{ss} + P_{buy}\right)$future days supply

Buy Quantity

The raw buy quantity, q, on a part is as follows:

$$\text{If}(\text{OH} + \text{OO} - \text{R}) > \text{OP} \qquad q = 0$$

$$\text{If}(\text{OH} + \text{OO} - \text{R}) \leq \text{OP} \qquad \text{q} = \text{OL} - (\text{OH} + \text{OO} - \text{R})$$

R is the requirements for the current day. In this special case where the buy from the supplier is at the start of every day, it is important to include R in the above computations.

The replenish quantity, Q, is computed as below.

$$\text{If } q \leq 0 \qquad Q = 0$$

$$\text{If } q > 0 \qquad \text{Q} = \text{ceiling}(q/M) \times M$$

For example, should $M = 12$ and $q = 17$, then $Q = 24$.

Example 5.5 Suppose part $h = 1$ with the following data:

$R_{\text{h}} = 100$
$R_{\text{t}} = R_{\text{ht}} = 100$ for t = 1–7
OH = 250
OO = 0
$M = 12$ pieces
$L = 2$ days
$P_{\text{ss}} = 1$ day
$P_{\text{buy}} = 4$ days.

At $t = 1$, the order point and order level are computed as follows:

$$\begin{aligned}\text{OP} &= (L{+}P_{\text{ss}}) = (2 + 1) \text{ days supply}\\ &= \text{R}_1 + \text{R}_2 + \text{R}_3 = 300\end{aligned}$$

$$\begin{aligned}\text{OL} &= \left(L + P_{\text{ss}} + P_{\text{buy}}\right) = (2 + 1 + 4) \text{ days supply}\\ &= (\text{R}_1 + \text{R}_2 + \text{R}_3 + \text{R}_4 + \text{R}_5 + \text{R}_6 + \text{R}_7) = 700\end{aligned}$$

Table 5.8 is a worksheet for the part and shows where $q_1 = 550$ at $t = 1$. Note the following calculations for day t:

$$\text{If}(\text{OH} + \text{OO})_{t-1} - \text{R}_t > \text{OP}_t \qquad \text{q}_t = 0$$

$$\text{If}(\text{OH} + \text{OO})_{t-1} - \text{R}_t \leq \text{OP}_t \qquad \text{q}_t = \text{OL}_t - \left[(\text{OH} + \text{OO})_{t-1} - \text{R}_t\right]$$

Using the multiple quantity, M, q_t is adjusted to Q_t, and the on-hand plus on-order quantity for time t is revised as below:

$$(\text{OH} + \text{OO})_t = (\text{OH} + \text{OO})_{t-1} - \text{R}_t + \text{Q}_t$$

In the example at $t = 1$, since

$$(\text{OH} + \text{OO})_0 - \text{R}_1 = 150 \leq \text{OP}_1 = 300, \text{q}_1 = 550.$$

Table 5.8 Inventory replenishment schedule for 7 days on part $h = 1$ with $M = 12$, $L = 2$, $P_{\text{ss}} = 1$ and $P_{\text{buy}} = 4$ days, OP = 300, OL = 700

t	1	2	3	4	5	6	7
R_t	100	100	100	100	100	100	100
$(\text{OH} + \text{OO})_{\text{t-1}}$	250	702	602	502	402	302	706
q_t	550	0	0	0	0	498	0
Q_t	552	0	0	0	0	504	0
$(\text{OH} + \text{OO})_{\text{t}}$	702	602	502	402	302	706	606

Table 5.9 Inventory replenishment worksheet for day 1 on the ten parts

h	M	R_h	OP_1	OL_1	OH_0	OO_0	q_1	Q_1
1	12	100	300	700	250	0	550	552
2	50	500	1500	3500	2100	0	0	0
3	100	100	300	700	120	400	0	0
4	30	100	300	700	80	100	620	630
5	24	200	600	1400	1100	0	0	0
6	1	100	300	700	600	100	0	0
7	50	100	300	700	80	0	720	750
8	40	400	1200	2800	2000	0	0	0
9	6	100	300	700	250	200	0	0
10	80	100	300	700	130	70	600	640

Because 550 is not a multiple of $M = 12$, the replenish quantity becomes $Q_1 = 552$. The ending inventory for day 1 becomes:

$$(\mathrm{OH} + \mathrm{OO})_1 = [250 - 100 + 552] = 702.$$

Note where the planned schedules out to day 7 are all zero except for day 6 where $Q_6 = 504$. For all the seven schedule requirements, the only one that is active is for day 1 that computes Q_1. The other schedules are for planning purposes only and are good information for the supplier to anticipate the replenishments that will be needed in the planning period.

Table 5.9 is a worksheet that summarizes the inventory replenishments for the ten parts of Table 5.2. Recall, the parameters are $P_{ss} = 1$ and $P_{buy} = 4$.

Summary

An important phase in assembly management is to ensure the stock required to carry on the assembly work is available at each station when needed. This requires the projection of the requirements by station on each part and/or component that is listed on the bill-of-material for the product. The chapter shows how to project the inventory requirements for single model lines, for mixed model make-to-stock lines and for mixed model make-to-order lines. The station requirements are used to compute the schedule of the incoming replenishments from the suppliers on each of the parts.

Chapter 6
Single Model Assembly

Introduction

This chapter concerns a plant that dedicates a line to produce a product that has no variation. This is called single model assembly. The planning methods that take place for this type of line are described in the context of an example. The example begins with the work elements, the element times, the predecessor elements, and the corresponding precedence diagram. The shift schedule quantity and shift time are needed to determine the number of operators to have on the line. The example shows how to assign the work elements to stations (line balancing) in order to obtain an even work load per station, as well as attain compliance with the precedence constraints. The example continues by showing how to measure the balance delay and efficiency ratio for the line. The bill-of-material data is used to identify the relation of parts to the work elements. This data allows the management to compute the requirement of parts for the shift schedule and for each station. The example also shows how the part requirements over the shift are replenished from the supplier. The replenishments could occur one time for the entire shift, or two or more times over the shift in the spirit of just-in-time deliveries.

Example 6.1 Consider a plant with a series of assembly lines that are available as needed for the variety of products that the firm produces. One of the products, called A, is under review. Top management is seeking $N = 45$ units of product A per day over the planning horizon. One shift is in use per day and the productive shift time is $T = 450$ min. The industrial engineers identify the work tasks that are needed to assemble one unit of product. These tasks are called work elements, or simply elements. The elements are labeled as $e = 1$ to Ne, where Ne is the number of elements for the product. Suppose product A has $Ne = 30$ elements and the task time of each, called the element time, t_e, is measured by the industrial engineers of the plant. The sum of the element times, $\sum t_e$, called the unit time for the product, is also tallied. Assume for product A, $\sum t_e = 38.2$ min.

N. T. Thomopoulos, *Assembly Line Planning and Control*,
DOI: 10.1007/978-3-319-01399-2_6,

Number of Operators

With T, $\sum t_e$, and N identified, the assembly management determines the number of operators, n, to assign on the line. The number of operators is computed by the following:

$$n' = N \times \sum t_e \Big/ T = 45 \times 38.2/450 = 3.82$$

Since, the number of operators must be an integer, $n' = 3.82$ is rounded up to $n = 4$.

Some Measures

It is now possible to measure the average time per operator on a unit. This is called the average operator time, denoted as $\bar{c}$, and is computed as below:

$$\bar{c} = \sum t_e/n = 38.2/4 = 9.55\,\text{min}$$

The cycle time, c', is the time between completed units coming off the line. This is measured as follows:

$$c' = T/N = 450/45 = 10.00\ \text{min.}$$

The potential efficiency of the line is measured by the balance delay, d, and alternatively by the efficiency ratio, E. In the example, these are as follows:

$$d = (c' - \bar{c})/c' = (10.00\text{–}9.55)/10.00 = 0.045$$

$$E = \bar{c}/c' = 9.55/10.00 = 0.955$$

The balance delay measures the portion of idle time on the line. Since $d = 0.045$, 4.5 % of the time on the line is idle.

The efficiency is another measure of the idle time on the line. With $E = 0.955$, 95.5 % of the time, the line is productive, and (1–0.955) = 0.045 gives 4.5 % as idle.

Predecessor Elements

When an element cannot be performed until another element has completed its task, the immediate prior element is called a predecessor element. Some elements have no immediate predecessors, and other have one or more predecessor elements. In the process or assigning elements to the n operators on a line, the

Table 6.1 List of elements, e, element times, t_e, and predecessor elements, p, for product A

e	t_e	p
1	1.2	
2	1.4	1
3	0.5	7
4	0.7	2
5	2.5	4
6	1.4	7
7	1.3	
8	0.9	4
9	0.4	3, 6
10	2.6	14
11	1.5	9
12	1.1	5
13	0.6	11, 14
14	1.7	8
15	2.1	13
16	1.4	15
17	1.5	6
18	1.1	15
19	2.3	13
20	0.9	19
21	0.5	17
22	0.2	17
23	0.6	17
24	1.1	20
25	1.5	23
26	0.6	23
27	0.9	26
28	2.1	22, 25
29	1.2	28
30	2.4	29

management must take into consideration the relation of the predecessor elements. To accommodate, a list of the elements along with their predecessor elements must be identified.

Continuing with the example, Table 6.1 contains a list of the elements, e, for product A, along with the element times, t_e. Recall, $Ne = 30$ is the number of elements, $\sum t_e = 38.2$ min is the unit time. Also, the predecessor element(s) are listed when they pertain. Note, for example, element $e = 9$ with element time $t_e = 0.4$ min has two predecessor elements, 3 and 6; element 7 has element time, $t_e = 1.3$ min and no predecessor elements, so forth.

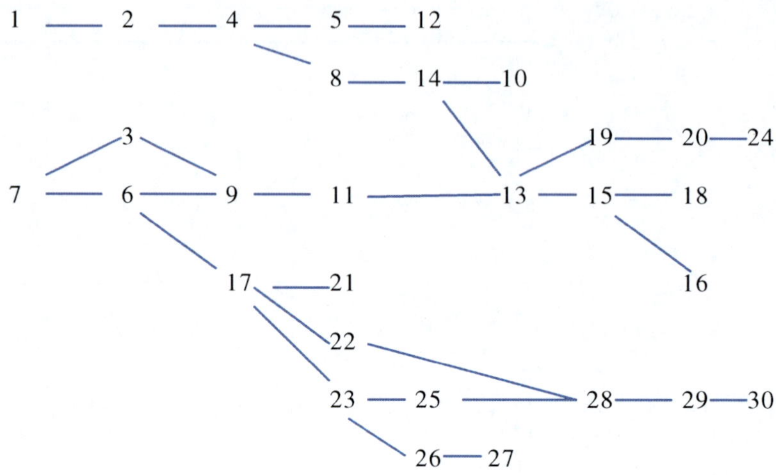

Fig. 6.1 Precedence diagram for the 30 element single model example

Precedence Diagram

The relation between all the elements and their predecessor elements are displayed in a precedence diagram as shown in Fig. 6.1. The sequence of elements moves from left to right. Note, elements 1 and 7 are on the left-hand side of the diagram and have no predecessor elements. Element 2 cannot begin until element 1 is completed. Elements 3 and 6 cannot begin till element 7 is completed. Also element 9, for instance, cannot begin until elements 3 and 6 are completed, and so on. The diagram depicts all the precedence relations as listed in Table 6.1.

Line Balancing

The assembly manager now has all the information needed to assign the elements to the operators on the line. The number or operators has been established at $n = 4$, and the average operation time is $\bar{c} = 9.55$. The elements are those listed in Table 6.1 along with their element times and predecessor elements. The task now is to assign the elements to the operators where the operator times are close to 9.55 min and where all the precedence constraints are satisfied.

One such assignment of the elements is listed in Table 6.2. The operators are denoted as $i = 1, 2, 3, 4$. Operator $i = 1$ is assigned elements $e = 1, 2, 4, 5, 8, 12$, and 14. The sum of the element times for the operator is $c_1 = 9.5$ min. None of the precedence constraints have been violated. In the same way, the table lists the elements assigned to operators $i = 2, 3$, and 4, with operator times of 9.8, 9.4, and 9.5 min, respectively. As before, all the precedence constraints are obeyed. Figure 6.2 shows how the element assignments are arranged by the four-operator stations.

Table 6.2 Station, i, assignment of elements, e, with element times, t_e, and station times, c_i

i	e	t_e	c_i
1	1	1.2	
	2	1.4	
	4	0.7	
	5	2.5	
	8	0.9	
	12	1.1	
	14	1.7	9.5
2	3	0.5	
	6	1.4	
	7	1.3	
	9	0.4	
	11	1.5	
	13	0.6	
	15	2.1	
	17	1.5	
	21	0.5	9.8
3	10	2.6	
	16	1.4	
	18	1.1	
	19	2.3	
	20	0.9	
	24	1.1	9.4
4	22	0.2	
	23	0.6	
	25	1.5	
	26	0.6	
	27	0.9	
	28	2.1	
	29	1.2	
	30	2.4	9.5

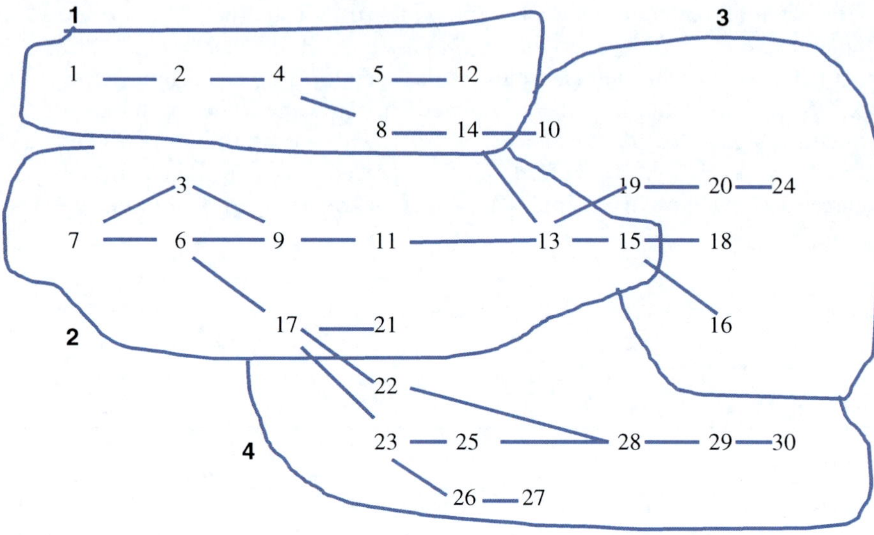

Fig. 6.2 Assignment of elements to the four operators

Effective Line Measures

The elements are now assigned to the station operators in a way where the operation times are close to the average of $\bar{c} = 9.55$ min and none of the precedence constraints are violated. The cycle time for the line becomes the following:

$$c = \max(c_1, c_2, c_3, c_4) = \max(9.5, 9.8, 9.4, 9.5) = 9.8 \text{ min}$$

The average operation time remains at $\bar{c} = 9.55$ min.

It is now possible to compute the effective measures of the balance delay and the efficiency ratio. The balance delay for the line is as below:

$$d = (c - \bar{c})/c = (9.80 - 9.55)/9.80 = 0.026$$

and the efficiency ratio is

$$E = \bar{c}/c = 9.55/9.80 = 0.974$$

Bill-of-Material Connection

In the typical bill-of-materials (bom) for a product, the list of parts (and components) needed on one unit of the product is identified. In assembly managements, it is also convenient to identify the element that first uses the part on the product.

Table 6.3 Part, h, bom number of units, b, and element, e

h	b	e
1	1	1
2	1	2
3	2	4
4	1	5
5	1	5
6	1	7
7	4	10
8	2	13
9	1	18
10	1	19
11	1	20
12	10	21
13	1	23
14	4	24
15	2	26
16	1	27

Table 6.3 is such a table for product A of the example. The columns list the parts, h, and the bom units of part, b, needed for each unit of product A. Another column identifies the element, e, that first uses the part. Altogether, the table shows sixteen parts are required in the assembly of product A.

This data is needed to determine and manage the inventory requirements so that the parts are available to comply with the shift schedule on the product. For example, one unit of part $h = 1$ is needed on one unit of product A and the first element to use the part is element $e = 1$. Two units of part $h = 3$ is needed for one unit of product A and the first element to use the part is element $e = 4$, so forth.

Shift Inventory Requirements

Recall for product A, the shift schedule calls for $N = 45$ units in the shift time of $T = 450$ min. Since the unit time to complete one unit of the product A is $\Sigma t_e = 38.2$ min, the number of operators (stations) needed is $n - 4$. One possible assignment of the elements to the four operators is shown in Table 6.2. This is one arrangement of line balancing. As shown earlier, the balance delay with this assignment is $d = 0.026$ and thereby the assignment of the elements is quite adequate.

The line balancing results in Table 6.2 and the bill-of-material data in Table 6.3 are now combined to determine the shift inventory requirements for each of the parts. Table 6.4 shows the shift inventory requirements, R_h, for each part h. Since one unit of part $h = 1$ is needed for each unit of product A, the shift inventory

Table 6.4 Station, i, with part, h, and part requirements, R_h, for each shift

i	h	R_h
1	1	45
1	2	45
1	3	90
1	4	45
1	5	45
2	6	45
2	8	90
2	12	450
3	7	180
3	9	45
3	10	45
3	11	45
3	14	180
4	13	45
4	15	90
4	16	45

needs of part $h = 1$ is $R_1 = b_h \times N = 1{\times}45 = 45$ units. In the same way, $R_3 = 2 \times 45 = 90$ units are needed to meet the shift requirements for the product.

Table 6.4 lists the part inventory results by operator, i. This data is needed by the management to identify where along the line to place the inventory for each of the parts. The table lists the operator and the number of pieces required. For example, operator $i = 1$, 45 pieces of parts $h = 1$, 2, 4, and 5 are needed, and 90 pieces of part $h = 3$.

Sequence of the Elements and Parts

Table 6.5 gives a possible sequence on how the elements will be performed from station 1 to station 4. The sequence also identifies the parts associated with each element, if any. When two or more units of the part are needed, the quantity is listed in parenthesis. So, for station, $i = 1$, the possible sequence of elements begins with $e = 1$, then is followed by $e = 2$, 4, 5, 12, 8, and 14. Part $h = 1$ is associated with element $e = 1$, and so forth. Note part $h = 3$ requires two units for each product A.

Just-in-Time Replenishments

Recall how Table 6.4 shows the shift requirements for each of the parts, labeled as R_h for part h. These quantities are again listed in Table 6.6 as R_1, signifying the part will be replenished once at the start of the entire shift. In the spirit of just-in-time,

Table 6.5 Potential sequence of elements, *e*, by operator, *i*. Also is the list of parts, *h*, with part quantity in parenthesis when more than one

i	*e*	*h*
1	1	1
	2	2
	4	3(2)
	5	4,5
	12	
	8	
	14	
2	7	6
	3	
	6	
	9	
	11	
	13	8(2)
	15	
	17	
	21	12(10)
3	10	7(4)
	19	10
	20	11
	24	14(4)
	18	9
	16	
4	22	
	23	13
	26	15(2)
	27	16
	25	
	28	
	29	
	30	

should the parts be replenished twice over the shift, the table shows the possible replenish quantities labeled as $R_{2.1}$ and $R_{2.2}$. In the event of replenishing the part four times over the shift, the replenish quantities for each of the parts are labeled as $R_{4.1}$, $R_{4.2}$, $R_{4.3}$, and $R_{4.4}$. Note for example in station $i = 1$, where part $h = 1$ would need $R_1 = 45$ units at the start of the shift if it is replenished only once. Should it be replenished two times in the shift, the replenish quantities are, 23 and 22. In the event the part is replenished four times during the shift, the replenish quantities are, 12, 11, 11, 11.

Table 6.6 List of parts, *h*, by station, *i*, with part replenishments when replenish, 1, 2, or 4 times during the shift

		Number replenishments per shift						
i	h	1	2		4			
		R1	R2.1	R2.2	R4.1	R4.2	R4.3	R2.4
1	1	45	23	22	12	11	11	11
	2	45	23	22	12	11	11	11
	3	90	45	45	23	22	23	22
	4	45	23	22	12	11	11	11
	5	45	23	22	12	11	11	11
2	6	45	23	22	12	11	11	11
	8	90	45	45	23	22	23	22
	12	450	225	225	113	112	113	112
3	7	180	90	90	45	45	45	45
	9	45	23	22	12	11	11	11
	10	45	23	22	12	11	11	11
	11	45	23	22	12	11	11	11
	14	180	90	90	45	45	45	45
4	13	45	23	22	12	11	11	11
	15	90	45	45	23	22	23	22
	16	45	23	22	12	11	11	11

Summary

This chapter is a review on the planning and control of a single model assembly line. It all begins with the work elements and the element times. The shift time and the shift schedule along with the sum of the element times are needed to determine the number of operators required on the line. After the number of operators is established, the elements are assigned to the operators in a way where the workload is evenly distributed to each of the operators. The efficiency of the line is measured by the balance delay and the efficiency ratio. Finally, the bill-of-material for the unit of product is used to determine the part and component inventory needs by station to satisfy the shift schedule requirements. In the spirit of just-in-time inventory, the station requirement needs can be fulfilled little by little. The example shows comparisons where the replenishments are once, twice, and four times over a shift.

Chapter 7
Mixed Model Make-to-Stock Assembly

Introduction

Mixed model make-to-stock assembly occurs when one line has two or more models in process at the same time. This chapter describes the planning methods that take place to control the operation of the line. The method is presented by an example or four models. The example begins with a listing of the work elements, the element times, the predecessor elements, and the element relation with the models called the usage index. The shift production time and the shift schedule for each model are also needed here. Next the stations are assigned the work elements where the assigned times over the shift are evenly distributed (line balancing). Each day, the sequence of the units down the line is generated in a way where the flow of work is as smooth as possible with minimum idleness and congestion at the stations. A make-to-stock sequencing algorithm (MSSA) demonstrates how the method works. Finally, the bill-of-material data is called to calculate the part requirements for each shift and for every station.

Firm ABC is a producer of appliances of all types, washing machines, dryers, refrigerators, and so on. Each type of appliance has various models. The firm has an assembly plant and a nearby distribution center (DC). The DC stocks the models of the appliances and awaits orders from the dealers located all over the country and beyond. Typically, three or more models of each type of appliance are in the firms line of products. The plant produces the appliances on assembly lines that are dedicated to each appliance type. All the models of each assembly type are produced together on a line in a mixed model way. Each model is a fixed design with no variation. The models have some variation among each other. The DC generates forecasts for each model of every appliance that spans the future months. The DC management wants to ensure an adequate supply on each model and computes the DC replenish needs for each of the models. The replenish quantities are delivered to the plant as requirements for the coming time periods. The models are produced on one or more mixed model lines with a make-to-stock arrangement. These are the type of products discussed in this chapter.

N. T. Thomopoulos, *Assembly Line Planning and Control*,
DOI: 10.1007/978-3-319-01399-2_7,

Example 7.1 Consider a mixed model make-to-stock assembly line with $N_e = 25$ work elements. The elements, e, are listed in Table 7.1 along with the elements times, t_e, and the immediate predecessor elements, p. Note where element $e = 1$ requires 2.4 min per unit of product and has no predecessor elements. Element $e = 3$ requires 1.9 min and has element $e = 1$ as a predecessor element; and element 8 that needs 2.2 min has two predecessors, $e = 2$ and 7, and so forth.

Precedence Diagram

The precedence diagram for the mixed model line is shown in Fig. 7.1. The sequence of elements is from left to right. Note, elements $e = 1$, 2, 4, and 5 are on the left-hand-side and they have no immediate predecessor elements. Element 3 has predecessor $e = 1$, and so forth.

Table 7.1 Element, e, element time, t_e, and predecessor elements, p

e	t_e	p
1	2.4	
2	3.2	
3	1.9	1
4	0.7	
5	1.9	
6	0.8	5
7	1.5	4
8	2.2	2,7
9	0.4	2,3
10	0.9	6
11	1.4	6
12	2.0	8
13	1.3	8,10
14	0.9	13
15	3.3	11
16	1.6	15
17	1.3	15
18	1.5	11
19	3.8	9
20	1.6	9
21	1.2	20
22	2.5	21
23	2.5	17,18
24	2.4	16
25	2.2	24

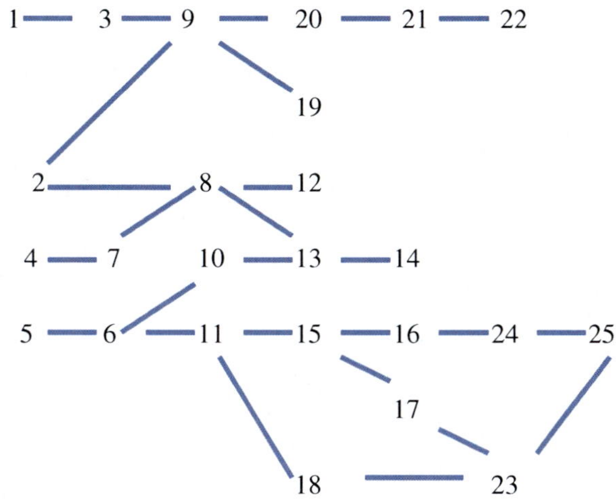

Fig. 7.1 Precedence diagram for the mixed model make-to-stock line

Mixed Model Shift Schedule

Suppose over the planning horizon, the daily shift schedule calls for $N = 105$ units of the four models ($j = 1, 2, 3, 4$). The typical mix of the schedule among the four models is listed in Table 7.2 where the model shift schedules are: $N_1 = 50$, $N_2 = 30$, $N_3 = 20$, and $N_4 = 5$ units. As described earlier, the number of units for the day is fairly constant over the planning horizon, but the mix of the models per day may vary slightly, depending on the demand mix of the models. To assign the elements to the stations, a typical mix of the schedule is used.

Shift Element Times

In order to assign the work elements to the stations along the line, the total time over a shift for each element, T_e, is needed. A worksheet for the computations is shown in Table 7.3. The table lists the elements, e, the element time, t_e, and the model usage per element denoted as u_{ej}. The model usage is defined as below:

Table 7.2 Models, j, and shift model schedule, N_j, for the mixed model make-to-stock line

j	N_j
1	50
2	30
3	20
4	5
Sum	105

Table 7.3 Elements, e, element time, t_e, model usage, u_{ej}, shift element model time, T_{ej} and shift element time, T_e

		u_{ej}				T_{ej}				
e	t	1	2	3	4	1	2	3	4	T_e
1	2.4	1	1	1	1	120	72	48	12	252
2	3.2	1	0	1	1	160	0	64	16	240
3	1.9	1	1	1	0	95	57	38	0	190
4	0.7	1	0	1	1	35	0	14	3.5	52.5
5	1.9	0	1	1	0	0	57	38	0	95
6	0.8	1	0	0	1	40	0	0	4	44
7	1.5	1	0	1	0	75	0	30	0	105
8	2.2	0	1	1	1	0	66	44	11	121
9	0.4	1	0	1	1	20	0	8	2	30
10	0.9	1	1	1	1	45	27	18	4.5	94.5
11	1.4	1	0	1	0	70	0	28	0	98
12	2	0	1	0	1	0	60	0	10	70
13	1.3	1	0	1	0	65	0	26	0	91
14	0.9	1	0	0	0	45	0	0	0	45
15	3.3	0	1	0	0	0	99	0	0	99
16	1.6	0	0	0	1	0	0	0	8	8
17	1.3	0	0	1	0	0	0	26	0	26
18	1.5	1	1	0	0	75	45	0	0	120
19	3.8	0	1	1	0	0	114	76	0	190
20	1.6	0	0	1	1	0	0	32	8	40
21	1.2	1	1	0	1	60	36	0	6	102
22	2.5	0	1	1	1	0	75	50	12.5	137.5
23	2.5	1	1	1	1	125	75	50	12.5	262.5
24	2.4	0	1	1	1	0	72	48	12	132
25	2.2	0	0	1	0	0	0	44	0	44
Sum										2689

u_{ej} = 1 if element e is used on model j
0 if element e is not used on model j

Note, in the table, where element $e = 1$ is used on all four models; element $e = 2$ is used on models 1, 3, 4, and so forth. The table also lists the shift model time for each element denoted as T_{ej}. This quantity is computed as shown below:

$$T_{ej} = N_j t_e \qquad \text{for } e = 1 \text{ to } N_e \text{ and } j = 1 \text{ to } Nj$$

For example, at $e = 1$ and $j = 1$, $T_{11} = N_1 \times t_1 = 50 \times 2.4 = 120$ min. So now the shift element times can be calculated. These are as follows:

$$T_e = \sum_j T_{ej} \qquad \text{for } e = 1 \text{ to } N_e$$

For element, $e = 1$, the shift time is $T_1 = (120 + 72 + 48 + 12) = 252$ min. The total time over all the elements for the shift is $\Sigma T_e = 2689$ min.

Number of Stations

A next step is to determine the number of operators, n, needed on the line. Because the shift time is $T = 450$ min and the total shift element time is $\Sigma T_e = 2689$ min, the number of operators required is computed as below:

$$n' = \sum T_e/T = 2689/450 = 5.97$$

Rounding up yields, $n = 6$ operators. So now the average shift time per operator becomes

$$\bar{T} = \sum T_e/n = 2689/6 = 448.2 \text{ min.}$$

Mixed Model Line Balancing

The next step is to assign the elements to the operators in a way where the shift time for each operator is fairly evenly distributed. To carry out this task, the data in Table 7.4 is needed. One result is shown in the table. The table is a list of the stations, i, the elements, e, assigned to each station, the element time, t_e, the usage by element and model, u_{ej}, and the shift element times, T_e. The line balancing goal is to assign the elements to the stations where the station shift times, T_i, are fairly close to average shift time, $\bar{T} = 448.2$ min and where all the precedence restrictions are in compliance. The table shows where the station shift times become the following: 442.0, 431.5, 448.5, 459.0, 469.5, and 438.5 min for stations 1 to 6, respectively. The maximum operator shift time is $T' = 469.5$ min for station 5, and so, the balance delay with this arrangement is computed by,

$$d = (T' - \bar{T})/T' = (469.5 - 448.2)/469.5 = 0.045$$

or 4.5 %.

The line balance results are depicted in Fig. 7.2, showing how the stations are aligned with the elements. Note elements 1 and 3 are assigned to station $i = 1$, elements 2, 4, 5, and 6 are with station 2, and so forth

Operator Model Times

Table 7.5 is a list of the operator times by model, denoted as c_{ij}. These are the measures of how much time operator i is assigned to each unit of model j in minutes. Operator $i = 1$ is assigned, 4.3, 2.4, 4.3, and 4.3 min to each unit of models, $j = 1$, 2, 3, and 4, respectively. Also, for $j = 1$, the assigned times are: 4.3, 4.7, 4.6, 3.7, 1.2, and 2.5 min, for operators 1 to 6, respectively. A perfect

Table 7.4 Station, i, elements, e, element time, t_e, model usage, u_{ej}, shift element time, T_e, and shift station time, T_i

			u_{ej}					
i	e	t_e	1	2	3	4	T_e	T_i
1	1	2.4	1	1	1	1	252	
1	3	1.9	1	0	1	1	190	442
2	2	3.2	1	1	1	0	240	
2	4	0.7	1	0	1	1	52.5	
2	5	1.9	0	1	1	0	95	
2	6	0.8	1	0	0	1	44	431.5
3	7	1.5	1	0	1	0	105	
3	8	2.2	0	1	1	1	121	
3	9	0.4	1	0	1	1	30	
3	10	0.9	1	1	1	1	94.5	
3	11	1.4	1	0	1	0	98	448.5
4	12	2	0	1	0	1	70	
4	13	1.3	1	0	1	0	91	
4	14	0.9	1	0	0	0	45	
4	15	3.3	0	1	0	0	99	
4	16	1.6	0	0	0	1	8	
4	17	1.3	0	0	1	0	26	
4	18	1.5	1	1	0	0	120	459
5	19	3.8	0	1	1	0	190	
5	20	1.6	0	0	1	1	40	
5	21	1.2	1	1	0	1	102	
5	22	2.5	0	1	1	1	137.5	469.5
6	23	2.5	1	1	1	1	262.5	
6	24	2.4	0	1	1	1	132	
6	25	2.2	0	0	1	0	44	438.5
Sum							2689	2689

assignment would occur when the model time is the same for each operator. However, such a result is most difficult to attain.

Make-to-Stock Sequencing Algorithm (MSSA)

Another important step in mixed model make-to-stock assembly is to determine the sequence of models down the line. If the mix of models is the same each day, the sequence determination is needed only once, but if the mix changes on a daily basis, a new sequence would be needed with each change. Typically, the mix does change each day, ever slightly, to account for changes in the demand mix on the models.

Below is an algorithm (MSSA) to determine the sequence. The algorithm seeks the sequence that gives the maximum space between two models of the same type. For instance, if 100 units are to be sequenced in the shift and 20 of them are for

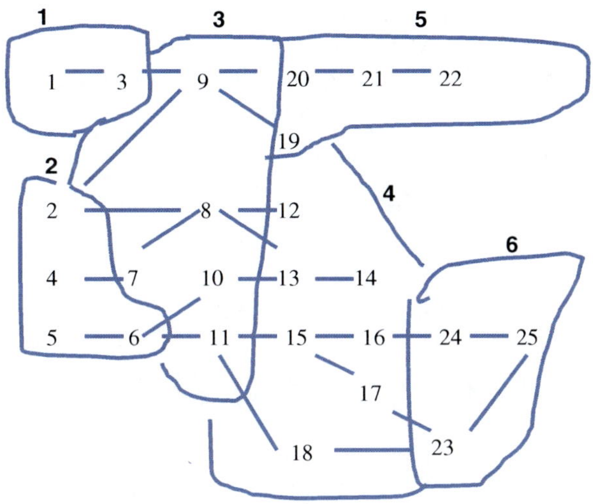

Fig. 7.2 Precedence diagram and element station assignments

Table 7.5 Station i and operator model times c_{ij} by model j

i	c_{ij} 1	2	3	4
1	4.3	2.4	4.3	4.3
2	4.7	5.1	5.8	4.7
3	4.6	1.7	4.6	2.1
4	3.7	6.8	2.6	3.6
5	1.2	7.5	7.9	5.3
6	2.5	4.9	7.1	4.9

model A and 33 are for model B, then the ideal sequence would be to place a model A as every fifth unit in the sequence and a model B every third unit. The data needed for the algorithm is merely the list of models and the shift schedule on each. This is the same data that appears in Table 7.2.

The algorithm is listed in Fig. 7.3 in a pseudo code way. The input data is the following:

j = model

Nj = number of models

N_j = shift schedule for model j

The output data is the following:

$j(i)$ = the sequence of models, $j(1)$ = first j in the sequence, $j(2)$ is the second, and so forth till $j(NN)$ is the final j in the sequence.

Another important measure, $a(j)$, is computed for every model j at each placement along the sequence. This is like a score for each model, and the model with the minimum score, $a(j)$, is selected at that placement.

In the example of this chapter, the number of models is four, and $N = 105$ units are to be produced where, the schedule calls for 50, 30, 20, and 5 units of models 1, 2, 3, and 4, respectively. The sequence results are listed in Table 7.6. The notation is seq as the sequence number that goes from 1 to 105. At each sequence

```
'INITIALIZE
NN = 0
For j = 1 to Nj
        N(j) = Nj
        NN = NN+N(j)
next j
'BEGIN
For j = 1 to Nj
        w(j) = NN/N(j)
        a(j) = w(j)/2
next j
For i = 1 to NN
        Min = a(1)
        For j = 1 to Nj
                If a(j)≤Min
                j(i) = j
                  Min = a(j)
                End if
        Next j
        For j = 1 to Nj
                if j <> j(i)
                   a(j) = a(j)
                end if
                If j =j(i)
                a(j)=a(j)+w(j)
                end if
        next j
next i
Return { j(i) i = 1 to NN }
'END
```

Fig. 7.3 Routine to sequence the models for a mixed model make-to-stock line (MSSA)

placement are the four computed values of $a(j)$. The model that is selected at that placement is the one with the minimum values of $a(j)$. The values of $a(j)$ are recalculated at each placement of the sequence. Note, at seq = 1, model $j = 1$ is selected since $a(1) = 1.05$ has the minimum value, and so forth.

Inventory Requirements

Another important step in the daily control of the assembly line is to determine the inventory requirements for each of the parts in the bill-of-material for the collection of models on the line.

Table 7.7 is a list of the parts, h, in the bill-of-material, and the bom quantity of units, b, needed for every unit of product. In addition, the table lists the first element, e that calls for use of the part. Further is the part model usage index, u_{ej},

Table 7.6 Sequences 1–105, seq, model score, *a*(*j*), and model *j*

seq	*a*(*j*) 1	2	3	4	*j*
1	1.05	1.75	2.63	10.5	1
2	3.15	1.75	2.63	10.5	2
3	3.15	5.25	2.63	10.5	3
4	3.15	5.25	7.88	10.5	1
5	5.25	5.25	7.88	10.5	2
6	5.25	8.75	7.88	10.5	1
7	7.35	8.75	7.88	10.5	1
8	9.45	8.75	7.88	10.5	3
9	9.45	8.75	13.13	10.5	2
10	9.45	12.25	13.13	10.5	1
11	11.55	12.25	13.13	10.5	4
12	11.55	12.25	13.13	31.5	1
13	13.65	12.25	13.13	31.5	2
14	13.65	15.75	13.13	31.5	3
15	13.65	15.75	18.38	31.5	1
16	15.75	15.75	18.38	31.5	1
17	17.85	15.75	18.38	31.5	2
18	17.85	19.25	18.38	31.5	1
19	19.95	19.25	18.38	31.5	3
20	19.95	19.25	23.63	31.5	2
21	19.95	22.75	23.63	31.5	1
22	22.05	22.75	23.63	31.5	1
23	24.15	22.75	23.63	31.5	2
24	24.15	26.25	23.63	31.5	3
25	24.15	26.25	28.88	31.5	1
26	26.25	26.25	28.88	31.5	2
27	26.25	29.75	28.88	31.5	1
28	28.35	29.75	28.88	31.5	1
29	30.45	29.75	28.88	31.5	3
30	30.45	29.75	34.13	31.5	2
31	30.45	33.25	34.13	31.5	1
32	32.55	33.25	34.13	31.5	4
33	32.55	33.25	34.13	52.5	1
34	34.65	33.25	34.13	52.5	2
35	34.65	36.75	34.13	52.5	3
36	34.65	36.75	39.38	52.5	1
37	36.75	36.75	39.38	52.5	2
38	36.75	40.25	39.38	52.5	1
39	38.85	40.25	39.38	52.5	1
40	40.95	40.25	39.38	52.5	3

(continued)

Table 7.6 (continued)

	a(j)				
seq	1	2	3	4	j
41	40.95	40.25	44.63	52.5	2
42	40.95	43.75	44.63	52.5	1
43	43.05	43.75	44.63	52.5	1
44	45.15	43.75	44.63	52.5	2
45	45.15	47.25	44.63	52.5	3
46	45.15	47.25	49.88	52.5	1
47	47.25	47.25	49.88	52.5	2
48	47.25	50.75	49.88	52.5	1
49	49.35	50.75	49.88	52.5	1
50	51.45	50.75	49.88	52.5	3
51	51.45	50.75	55.13	52.5	2
52	51.45	54.25	55.13	52.5	1
53	53.55	54.25	55.13	52.5	4
54	53.55	54.25	55.13	73.5	1
55	55.65	54.25	55.13	73.5	2
56	55.65	57.75	55.13	73.5	3
57	55.65	57.75	60.38	73.5	1
58	57.75	57.75	60.38	73.5	2
59	57.75	61.25	60.38	73.5	1
60	59.85	61.25	60.38	73.5	1
61	61.95	61.25	60.38	73.5	3
62	61.95	61.25	65.63	73.5	2
63	61.95	64.75	65.63	73.5	1
64	64.05	64.75	65.63	73.5	1
65	66.15	64.75	65.63	73.5	2
66	66.15	68.25	65.63	73.5	3
67	66.15	68.25	70.88	73.5	1
68	68.25	68.25	70.88	73.5	2
69	68.25	71.75	70.88	73.5	1
70	70.35	71.75	70.88	73.5	1
71	72.45	71.75	70.88	73.5	3
72	72.45	71.75	76.13	73.5	2
73	72.45	75.25	76.13	73.5	1
74	74.55	75.25	76.13	73.5	4
75	74.55	75.25	76.13	94.5	1
76	76.65	75.25	76.13	94.5	2
77	76.65	78.75	76.13	94.5	3
78	76.65	78.75	81.38	94.5	1
79	78.75	78.75	81.38	94.5	1
80	80.85	78.75	81.38	94.5	2

(continued)

Table 7.6 (continued)

seq	a(j) 1	2	3	4	j
81	80.85	82.25	81.38	94.5	1
82	82.95	82.25	81.38	94.5	3
83	82.95	82.25	86.63	94.5	2
84	82.95	85.75	86.63	94.5	1
85	85.05	85.75	86.63	94.5	1
86	87.15	85.75	86.63	94.5	2
87	87.15	89.25	86.63	94.5	3
88	87.15	89.25	91.88	94.5	1
89	89.25	89.25	91.88	94.5	1
90	91.35	89.25	91.88	94.5	2
91	91.35	92.75	91.88	94.5	1
92	93.45	92.75	91.88	94.5	3
93	93.45	92.75	97.13	94.5	2
94	93.45	96.25	97.13	94.5	1
95	95.55	96.25	97.13	94.5	4
96	95.55	96.25	97.13	115.5	1
97	97.65	96.25	97.13	115.5	2
98	97.65	99.75	97.13	115.5	3
99	97.65	99.75	102.38	115.5	1
100	99.75	99.75	102.38	115.5	1
101	101.85	99.75	102.38	115.5	2
102	101.85	103.25	102.38	115.5	1
103	103.95	103.25	102.38	115.5	3
104	103.95	103.25	107.63	115.5	2
105	103.95	106.75	107.63	115.5	1

Table 7.7 Parts, h, bom number per unit, b_h, element, e, element usage, u_{ej}, shift element usage, N_e, and shift part requirement, R_h

h	b_h	e	u_{ej} 1	2	3	4	N_e	R_h
1	2	1	1	1	1	1	105	210
2	1	2	1	1	1	0	100	100
3	1	4	1	0	1	1	75	75
4	1	5	0	1	1	0	50	50
5	4	6	0	1	1	0	50	200
6	2	7	1	0	1	0	70	140
7	1	10	1	1	1	1	105	105
8	1	13	1	0	1	0	70	70
9	1	18	1	1	0	0	80	80
10	10	19	0	1	1	0	50	500
11	1	20	0	0	1	1	25	25
12	4	21	1	1	0	1	85	340
13	2	23	1	1	1	1	105	210
14	1	24	0	1	1	1	55	55

Table 7.8 Station, i, element, e, part, h, and shift part requirement, R_h

i	e	h	R_h
1	1	1	210
2	2	2	100
	4	3	75
	5	4	50
	6	5	200
3	7	6	140
	10	7	105
4	13	8	70
	18	9	80
5	19	10	500
	20	11	25
	21	12	340
6	23	13	210
	24	14	55

with values of 0 and 1. The value is 0 when model j does not apply element e, and is set to 1 when element e is used on model j. N_e is the frequency of times over the shift that element e is in use. Finally, R_h is the shift requirement for part h.

Below shows how the measures of N_e and R_h are computed:

$$N_e = \sum\nolimits_j u_{ej} N_j \qquad \text{for } e = 1 \text{ to } N_e$$

$$R_h = N_e b_h \qquad \text{for } h = 1 \text{ to } N_h$$

Note for part $h = 1$, $N_1 = 105$ since element $e = 1$ is used on all four models, and R_1 is 210 since two units of $h = 1$ are needed on every unit of the product. At $h = 2$, $N_2 = 100$ since the element $e = 2$ is used on all models but $j = 4$. The part $h = 2$ requirements is $R_2 = 100$ since one unit of the part is placed on each unit of product.

The line balance results are now used to identify where on the line the parts are needed. The results are listed in Table 7.8. Because element $e = 1$ is assigned to operator $i = 1$, the associated part, $h = 1$ is also associated with operator 1. The shift requirements for the part is listed as $R_1 = 210$, and so forth.

Summary

In a mixed model make-to-stock line, the shift schedule quantity by model and the shift production time are used along with the unit times by model to determine the number of operators needed on the line. A mixed model precedence diagram is applied to assign the elements to each of the operators in a way where the shift work time per operator is evenly distributed. Next, a sequencing algorithm is used each day to generate the order in sending the models down the line. The bill-of-material for each model is then used to determine the part requirements for the shift and for each station.

Chapter 8
Mixed Model Make-to-Order Assembly

Introduction

Make to order assembly occurs when the customers specify the features and options for each unit they buy. The methods to control the operation of this type of assembly line are described by an example. The example begins with the work elements, element times, predecessor elements and any associated features and options. The probability of options by feature is used to project the number of options and features by shift over the planning horizon. The shift time and the number of units to build over the shift are then used to estimate the element times over the shift. With this information, the assignment of elements to the stations (line balancing) is carried out. An order board contains all the current customer orders, called jobs, with the exact feature and option combinations, as well as the due dates. The management selects the jobs for a forthcoming daily shift. The next decision is how to sequence the jobs down the line. A make-to-order sequencing algorithm (MOSA) is introduced and is demonstrated with a shift schedule of fifty jobs. Sometimes, before the sequence date, a job that is scheduled on a sequence has to be removed and replaced by another job taken from a pool of candidate jobs. A make-to-order replacement algorithm (MORA) is presented to select the substitute job, from the pool of jobs, for this purpose, and is demonstrated in the example. Finally, the bill-of-material for the parts enters as data and is used to determine the part requirements for each shift and station.

Make-to-Order Assembly

Truck manufacturers typically use make-to-order assembly. The number of units going down the line each day in a shift is a constant that is established for all the days over the planning horizon. The units are expensive and are individually produced to the specifications of the customers. Each truck comes with a list of features and every feature has a variety of options. The customer selects the

N. T. Thomopoulos, *Assembly Line Planning and Control*,
DOI: 10.1007/978-3-319-01399-2_8,

options wanted for every feature, and sometimes the customer bypasses a particular feature altogether, called a null option. The combination of options and features make each unit a unique model. Often, for an assembly shift, no two units going down the line are the same. The units are called jobs and the job has a list of specifications by feature and option. Every job also has a due date as promised to the customer. The jobs are assigned to a shift about a week prior to their assembly down the line when all the parts and components needed for the job are available and as close to the job due date as possible. After all the jobs are selected for a shift, the sequence of the jobs down the line is decided and this takes place several days prior to assembly. The sequencing seeks to spread the jobs down the line in a manner where the options by feature are as far apart as possible. This is needed since some of options on a feature take longer to process than others. An algorithm to accomplish the sequence is provided in the chapter. With the sequence available several days in advance of the shift, the management can arrange to have all the particular parts and components by job ready at the assigned stations along the line, and in that way the assembly process can take place in a smoothed manner.

Example 8.1 Assume a make-to-order assembly line where the shift time is $T = 450$ min and the shift schedule is $N = 50$ units. Each unit of the shift has the choice of four features *f*, and the features have a variety of options *k*. Altogether there is $Ne = 28$ work elements and some of the vital data of each are itemized in Table 8.1.

Each element is listed along with it's work element time t_e. For those with one or more predecessor elements, the immediate predecessor elements *p*, are included in the table. For each element associated with a feature, the particular feature *f*, is noted. Because the element time can vary by the option of a feature, those elements with features have an element time that represents the weighted average time over the options. This average is used subsequently in the line balancing stage. Note, element $e = 6$ is associated with feature $f = 3$; element $e = 12$ with feature 1, and so forth.

Precedence Diagram

Figure 8.1 is the precedence diagram for the elements in Table 8.1. Elements $e = 3$ and 6 have no predecessor elements. Element $e = 2$ cannot begin until element $e = 3$ is completed, and so forth.

Features and Options

Table 8.2 shows all the combination of features and options and a probability estimate for each combination. Note there are $Nf = 4$ features and for each feature *f*, there are Nk_f non-null options. Option $k = 0$ is when the customer chooses to not

Table 8.1 Work elements e, element times t_e, predecessor elements p, and features f

e	t_e	p	f
1	0.8	6	
2	2.8	3	
3	1.6		
4	1.2	6	
5	1.1	6	
6	2.5		3
7	0.5	1	
8	0.7	7	
9	0.7	5	
10	0.6	6	
11	0.8	10	
12	2.4	19	1
13	1.2	11	
14	1.9	15	
15	0.7	8	
16	1.9	12	2
17	0.8	13	
18	1.5	17	
19	2.2	2	4
20	0.4	12	
21	0.9	17	
22	1.4	14,16	
23	1.2	21	
24	1.3	23	
25	0.9	23	
26	2.7	14	
27	1.6	26	
28	1.3	27	

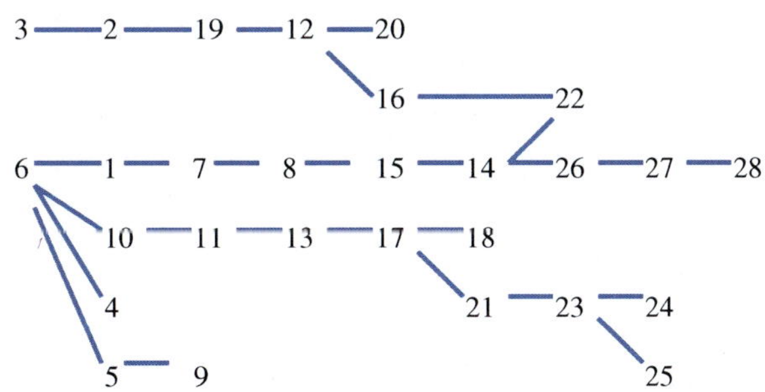

Fig. 8.1 Precedence diagram for the 28 elements

apply any of the options for the feature, called a null option. So, for feature $f = 1$, $Nk_1 = 5$ indicating there are five options for feature $f = 1$. There also is another option of $k = 0$, which is to not use any of the other five options. Note feature $f = 2$ has three options, $f = 3$ has one, and $f = 4$ has four options. The estimate of

Table 8.2 Probabilities $P(f, k)$, of option k, by feature f

f\k	0	1	2	3	4	5
1	0.10	0.20	0.20	0.25	0.15	0.10
2	0.20	0.30	0.40	0.10		
3	0.50	0.50				
4	0.10	0.30	0.30	0.20	0.10	

the probabilities will sum to one for each feature. Note for feature $f = 1$, 10 % of the customers choose none of the five options, 20 % chose option 1, and so forth. The sum of the probabilities for feature $f = 1$ is 1.00 (100 %).

Element Shift Times

The data from Table 8.2 is now used to estimate how often each of the elements will be used during an average shift, and also how much of the element time is needed in a shift duration. For those elements that do not need a feature, they will be used on every unit going down the line in a shift. For the elements with a feature, not all units will apply the element. Recall 10 % of the elements with feature $f = 1$ will not use any of the options for the feature, and so, only 90 % of the units going down the line will apply any element that is associated with feature $f = 1$.

Table 8.3 is a list of the elements again with some new data. The table lists the elements, e, the element time, t, and the feature, f, if any. This is the same data as provided earlier in Table 8.1. Recall, for the example, the shift schedule calls for $N = 50$ units. In this table, N_e designates the frequency of use for each element e over a shift. All the elements that do not have an associated feature, f, are used on all the units coming down the line and therefore, $N_e = N = 50$. Those elements that are aligned with a feature, are not used on all the units coming down the line. Note, element $e = 6$, is aligned with feature $f = 3$, and the probability of the null option for the feature is 0.50, and thereby, $N_e = (1 - 0.50) \times N = 0.50 \times 50 = 25$. In the same way N_e is computed for the other elements aligned with a feature, $e = 12, 16$, and 19. The shift time for each of the elements is denoted as T_e and is obtained by the relation,

$$T_e = N_e \times t_e \quad e = 1 \text{ to } Ne$$

Note, for $e = 1$, $T_1 = 50 \times 0.8 = 40$ min, for $e = 6$, $T_6 = 25 \times 2.5 = 62.5$ min, and so forth. At the bottom of the table is the sum of all the shift element times, where,

$$\sum T_e = 1776 \, \text{min}.$$

Table 8.3 Elements e, element time t_e, features f, shift element frequency N_e, and shift element time T_e

e	t_e	f	N_e	T_e
1	0.8		50	40
2	2.8		50	140
3	1.6		50	80
4	1.2		50	60
5	1.1		50	55
6	2.5	3	25	62.5
7	0.5		50	25
8	0.7		50	35
9	0.7		50	35
10	0.6		50	30
11	0.8		50	40
12	2.4	1	45	108
13	1.2		50	60
14	1.9		50	95
15	0.7		50	35
16	1.9	2	40	76
17	0.8		50	40
18	1.5		50	75
19	2.2	4	45	99
20	0.4		50	20
21	0.9		50	45
22	1.4		50	70
23	1.2		50	60
24	1.3		50	65
25	0.9		50	45
26	2.7		50	135
27	1.6		50	80
28	1.3		50	65
Sum				1776

Number of Stations

Recall the shift productive time is set at $T = 450$ min. Using T and the shift element time, $\sum T_e = 1776$, it is now possible to determine the number of stations needed on the line. This is by

$$n' = \sum T_e \big/ T = 1776/450 = 3.95$$

Rounding up gives, $n = 4$ stations.

Efficiency

The average shift time by station would then become,

$$\bar{T} = \sum T_e/n = 1776/4 = 444\,\text{min}.$$

This is slightly below, the shift time of $T = 450$ min. So now, the efficiency on the line becomes,

$$E = \bar{T}/T = 444/450 = 0.987$$

or 98.7 %.

Line Balancing

Assigning the work elements to the operators (line balancing) is needed only once, as long as the number of jobs per shift is constant and also as the combination of features and options is relatively the same. The planning horizon has set the shift schedule at $N = 50$ units. It further has estimated the feature option probabilities as those of Table 8.2. This allows the management to assign the elements to the line in a way where the typical daily workload by station is estimated accordingly. The management is fully aware that the daily workload schedule will vary according to the daily schedule of jobs and their specifications of features and options.

Now with $n = 4$ stations established for the line, the next step is to assign the work elements to the stations. This is the line balance phase. The data from Tables 8.1 and 8.3 are used. Table 8.1 includes the predecessor elements and Table 8.3 the shift element times for every element e. The goal is to assign the elements to each of the four stations in a way where the station shift times T_i are close to the average shift time $\bar{T} = 444$ min, and also where all the precedence constraints are satisfied properly.

One solution is listed in Table 8.4 where the station shift times are: 443, 447, 441 and 445 min for stations, 1–4, respectively. This seems like a very efficient assignment. Note, the total time over the four stations is $\sum T_i = 1776$ min. The assignment of the elements to the stations is depicted in Fig. 8.2.

The Shift Job Schedule

For every shift along the planning horizon, a schedule of the jobs to produce during the shift is gathered. In the example, the shift scheduled calls for $N = 50$ units, and thereby 50 jobs are chosen from the order board for assembly on a shift.

Table 8.4 Station i, element e, element time t_e, feature f, shift element time T_e, and shift station time T_i

i	e	t_e	f	T_e	T_i
1	1	0.8		40	
	4	1.2		60	
	5	1.1		55	
	6	2.5	3	62.5	
	7	0.5		25	
	8	0.7		35	
	9	0.7		35	
	10	0.6		30	
	11	0.8		40	
	13	1.2		60	443
2	2	2.8		140	
	3	1.6		80	
	12	2.4	1	108	
	19	2.2	4	99	
	20	0.4		20	447
3	15	0.7		35	
	16	1.9	2	76	
	17	0.8		40	
	18	1.5		75	
	21	0.9		45	
	23	1.2		60	
	24	1.3		65	
	25	0.9		45	441
4	14	1.9		95	
	22	1.4		70	
	26	2.7		135	
	27	1.6		80	
	28	1.3		65	445
Sum					1776

The jobs are chosen where the variety of parts and components are available to complete the assembly, and also where the assembly date is compatible with the due date of each job.

Table 8.5 is a list of the 50 jobs scheduled for a particular shift. The table lists the feature option combination for each of the jobs. For simplicity, the jobs are labeled as $j = 1$ to 50. Job $j = 1$, for example, has options 4, 2, 1, and 1 for features, 1–4, respectively. Note job $j = 2$ has the null option $k = 0$ for feature $f = 3$, indicating the customer does not want feature $f = 3$.

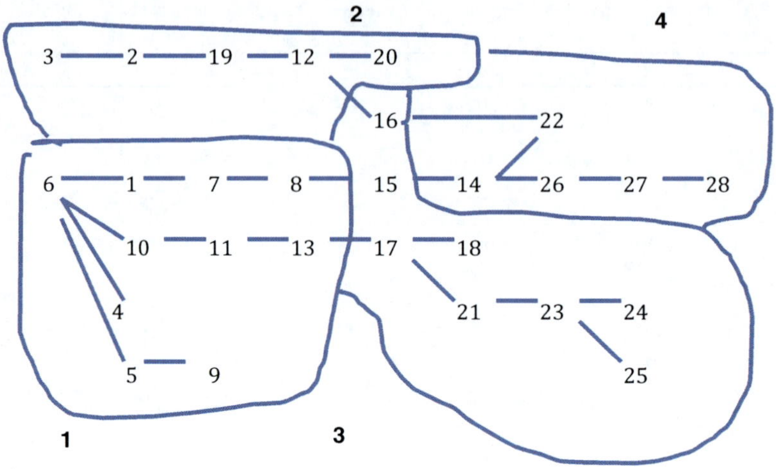

Fig. 8.2 Line balancing assignments of elements to stations

Shift Count of Feature Options

The 50 jobs assigned for the shift are now tallied to count the frequency of feature and option combinations, denoted as $N(f, k)$. The frequency is shown in Table 8.6. Note for feature, $f = 1$, 5 of the 50 jobs use the null option, 10 use option $k = 1$, and so forth. The reader should realize, the frequencies of feature options will vary for each of the days along the planning horizon. The data from this table is used in the sequencing phase on the line. Since the table frequencies change daily, the sequencing of the jobs down the line also varies each day.

Make-to-Order Sequencing Algorithm

The goal of the make-to-order sequencing algorithm (MOSA) is to arrange the jobs down the line in a way where the feature-options are spread apart as much as possible. Should option 1 of a feature be selected on 20 % of the jobs, say, the ideal sequence will schedule option 1 (of the feature) on every fifth unit going down the line. As the number of features goes up and the variety of options increases, the sequencing process becomes more complicated. Below describes the algorithm in pseudo code that achieves the goal stated.

The notation for this section is listed below:

I = sequence index (seq)
j = job index
f = feature index

Table 8.5 Jobs j, with features f and options k

f	1	2	3	4
	k			
j				
1	4	2	1	1
2	2	2	0	3
3	4	2	0	2
4	4	2	0	4
5	4	0	1	1
6	3	2	0	2
7	2	1	1	3
8	1	1	1	3
9	3	3	1	1
10	3	3	0	2
11	1	3	1	0
12	3	0	0	3
13	1	0	0	2
14	2	3	1	2
15	1	0	0	3
16	2	1	1	1
17	3	1	0	2
18	0	1	1	1
19	4	1	0	4
20	3	2	0	0
21	3	2	1	3
22	0	2	1	2
23	3	2	1	2
24	2	1	0	0
25	1	3	0	2
26	2	1	0	2
27	1	2	1	1
28	5	2	1	2
29	1	2	0	3
30	3	2	0	1
31	0	0	0	1
32	0	2	1	2
33	4	0	0	3
34	2	1	0	3
35	5	2	0	3
36	2	1	0	1
37	2	0	1	1
38	5	0	0	1
39	4	2	0	3
40	1	0	0	1

(continued)

Table 8.5 (continued)

f	1	2	3	4
		k		
41	4	1	0	2
42	1	1	0	2
43	1	3	1	3
44	5	1	1	0
45	3	1	1	0
46	5	2	0	1
47	2	1	1	1
48	0	1	1	4
49	3	1	1	3
50	3	2	1	1

Table 8.6 Shift frequency $N(f, k)$, of option k, by feature f, for a particular day

f/k	0	1	2	3	4	5
1	5	10	10	12	8	5
2	9	17	18	6		
3	27	23				
4	5	15	14	13	3	

k = option index
$k(j, f)$ = option for feature f of job j
Nj = number of jobs
Nf = number of features
$N(f, k)$ = shift number of option k for feature f
$x(f, k)$ = a measure of last sequence of option k for feature f
$y(j, f)$ = a measure for every job j not yet assigned and feature f
$Y(j)$ = measure for job j at sequence I
$J(I)$ = job j selected at sequence I

The sequencing algorithm is described in the six steps given below.

Step 1. Perform steps 2–6, below, for each sequence index I where $I = 1$ to Nj.

Step 2. Let jo = the job assigned in the prior sequence, (I−1), and let ko represent $k(jo,f)$ = the option of feature f that is associated with job jo.

Step 3. Keep a measure on the last use on each feature-option as below:

$x(f, k)$ = a measure on the last sequence assignment for option k of feature f. After every job assignment, each of the feature measures, $x(f, ko)$, are updated in the following way.

$$x(f,ko) = [x(f,ko) + Nj\,/\,N(f,ko)] \text{ where } ko = k(jo,f).$$

Step 4. To select the next job for sequence index I, the following measures, $y(j, f)$, are obtained for each of the features f and for every job j that is not yet assigned. This measure gives–ideally–how far in advance to place the next unit with feature f of job j. The measure for job j and feature f is obtained as below.

$$y(j,f) = [x(f,k) + Nj/N(f,k)] \text{ and } k = k(j,f).$$

Step 5. For each job j that is not yet assigned, the sum of $y(j, f)$ over all features is computed. This sum, $Y(j)$, gives a measure for job j. This measure can then be compared to all unassigned jobs in a relative way. The sum is computed as below:

$$Y(j) = \sum\nolimits_f [y(j,f)]$$

Step 6. With the above sums, $Y(j)$, determined for each unassigned job, the best job to assign for the current sequence index I, is the job with the minimum such measure. That is, choose job jo as follows:

$Y(jo) = \min\ \{Y(j)$ for all j not yet assigned$\}$.

$J(I) = jo$ is the job chosen at sequence I of the schedule.

A difficulty with the above algorithm concerns the initializing stage for the sequence since no prior values at $I = 0$ of $x(f, k)$ are available to begin the algorithm at $I = 1$. To overcome, and assuming, the sequencing is performed each day, the measures for each feature-option, $x(f, k)$, should be saved from the last unit of the prior days' sequence; and that last unit should give the measure for $I = 0$ for the next days' run. In the event the sequence data from the prior day is not available, the initial set of measures can be computed as below: $x(f, k) = [0.5 \times Nj/N(f, k)]$. This measure is needed for every combination of feature and option.

Make-to-Order Sequence

The sequence for the shift is summarized in Table 8.7. The table columns are $I =$ sequence number, $J(I) = j =$ job, $f =$ 1–4 with features and the corresponding options k. Finally is a measure, $Y(j)$, from the sequence algorithm that is computed for each unassigned job in the schedule. The algorithm selects the job with the minimum $Y(j)$ value. At the outset when $I = 1$, the job with the minimum measure ($Y(j) = 18.10$) is $j = 30$, and thereby job $J(1) = j = 30$ is selected as the first job in the sequence.

The goal of the algorithm is to spread the feature options as far apart as possible. For example, option $k = 3$ of feature $f = 1$ is assigned on the sequence as $I = 1, 5, 9, 14$, and so on. This spaces option 3 of feature 1 as far apart as feasible. The algorithm does the same on all options and features.

Table 8.7 Sequence results: spot *I* has job *J(I)* with options *k* for feature *f*, and sequence measure *Y(j)*

I	j = J(I)	f	k				Y(j)
			1	2	3	4	
1	30		3	2	0	1	18.10
2	8		1	1	1	3	20.94
3	26		2	1	0	2	23.91
4	5		4	0	1	1	28.70
5	6		3	2	0	2	31.84
6	29		1	2	0	3	39.24
7	18		0	1	1	1	39.57
8	24		2	1	0	0	44.99
9	23		3	2	1	2	49.36
10	33		4	0	0	3	51.30
11	14		2	3	1	2	58.02
12	46		5	2	0	1	58.24
13	15		1	0	0	3	66.28
14	45		3	1	1	0	69.05
15	32		0	2	1	2	74.00
16	36		2	1	0	1	76.61
17	39		4	2	0	3	82.37
18	9		3	3	1	1	83.89
19	42		1	1	0	2	88.14
20	38		5	0	0	1	94.43
21	21		3	2	1	3	96.34
22	19		4	1	0	4	102.18
23	27		1	2	1	1	105.04
24	7		2	1	1	3	109.28
25	13		1	0	0	2	112.98
26	22		0	2	1	2	116.69

(continued)

Table 8.7 (continued)

I	j = J(I)	f	k				Y(j)
			1	2	3	4	
27	20		3	2	0	0	120.96
28	47		2	1	1	1	124.39
29	10		3	3	0	2	128.13
30	35		5	2	0	3	133.88
31	37		2	0	1	1	137.34
32	41		4	1	0	2	139.02
33	43		1	3	1	3	145.22
34	50		3	2	1	1	151.28
35	34		2	1	0	3	154.82
36	31		0	0	0	1	157.58
37	3		4	2	0	2	160.85
38	44		5	1	1	0	162.73
39	49		3	1	1	3	170.83
40	4		4	2	0	4	172.31
41	40		1	0	0	1	174.52
42	25		1	3	0	2	182.40
43	16		2	1	1	1	183.80
44	17		3	1	0	2	190.94
45	2		2	2	0	3	194.55
46	1		4	2	1	1	197.95
47	11		1	3	1	0	203.38
48	12		3	0	0	3	203.99
49	28		5	2	1	2	207.07
50	48		0	1	1	4	210.86

So for the shift of the day, the sequence of jobs is established. Typically, the sequence is generated several days in advance of the actual shift date. This allows the management to arrange to have the parts and components available at each station so that the assembly of the 50 jobs can flow smoothly for the shift.

Job Replacements

As stated earlier, the sequence for the shift is established several days ahead of the actual shift day. This allows the management to have the parts and components stock available as needed over the shift sequence. But a problem sometimes occurs with the schedule of jobs. This is when one or more of the jobs in the schedule has to be replaced for one reason or another. Could be for a job when the parts are not available to finish the assembly, or the customer (of the job) changes the specifications.

When a job is dropped, another job must take its place in the exact spot of the sequence. It is otherwise too disruptive to alter the sequence of the remaining jobs of the schedule. A list of the candidate replacement jobs is needed. From this list, the management must select one of the candidate jobs to replace the dropped job.

Make-to-Order Replacement Algorithm

The goal of make-to-order job replacement algorithm, MORA, is the find a job from the candidate of jobs to replace the dropped job j. The replaced job j should be as similar to the dropped job as possible and as dissimilar to the jobs before and after. As each candidate job j is analyzed, points $B(j)$ are tallied. $B(j)$ is a tally for each candidate job j on how close the job is to the replaced job, and how dissimilar it is to the jobs before and after the replaced job. The candidate job with the maximum points, $B(j)$, is selected as the replaced job and will be inserted in the same spot in the sequence as the dropped job.

The algorithm is listed below in a pseudo code manner.

```
Notation:
j1 = dropped job
j0 = job before j1 in sequence
j2 = job after j1 in sequence
f = features
Nf = number of features
k = option of feature
j = j3 = candidate jobs
k(j,f) = option k of feature f for job j
Nj3 = number candidate jobs

For j = 1 to Nj3
  j= j3
  B(j) = 0
 For f = 1 to Nf
  k0=k(j0,f)
  k1=k(j1,f)
  k2=k(j2,f)


  k3=k(j3,f)
 if k3=k0      then   B(j) = B(j) - 1
 if k3=k1      then   B(j) = B(j) + 1
 if k3=k2      then   B(j) = B(j) - 1
  Next f
 Next j
 Replace job = j with Min{ B(j) }
```

Drop Job 23

In the example, suppose job $j = 23$ is dropped. Note this is the ninth job in the shift sequence, and the options are $k = 3, 2, 1, 2$ for features $f = 1$–4, respectively. It now behooves the management to gather a list of the jobs that are eligible to replace the dropped job. Suppose $Nj3 = 20$ candidates are available and are the jobs listed in Table 8.8. Ideally, a candidate job with the same options (3, 2, 1, 2) is sought. The table lists the candidate jobs (51–70)

Drop Job 23 and Replace with Job 66

Table 8.9 lists the output results for each of the candidate jobs (51–70) from the make-to-order job replacement algorithm. Each candidate has the points,$B(j)$, computed in the algorithm. The job with the most points is chosen as the replacement. This is the job that is closest to the dropped job in the combination of

Table 8.8 Candidate jobs *j*, with features *f*, and options *k*

j	f 1 k	2 –	3 –	4 –
51	2	1	1	1
52	2	2	1	2
53	2	2	0	2
54	2	2	1	1
55	2	1	0	2
56	1	2	0	4
57	2	1	1	3
58	3	2	0	3
59	1	3	0	0
60	4	1	0	0
61	1	1	0	3
62	3	2	0	2
63	4	2	0	4
64	4	0	1	3
65	0	1	0	1
66	5	2	1	2
67	4	1	0	2
68	5	0	1	1
69	1	1	0	2
70	5	2	0	4

features and options, and also remains dissimilar to the jobs before and after in the sequence. Note the following option indices for the dropped and replaced jobs and also for the jobs before and after. The option indices are for features 1–4.

	job before:	2, 1, 0, 0
Dropped job:	job j = 23:	3, 2, 1, 2
	job after:	4, 0, 0, 3
	job before:	2, 1, 0, 0
Replaced job:	job j = 66:	5, 2, 1, 2
	job after:	4, 0, 0, 3

With the results of the algorithm, job $j = 23$ is replaced in the ninth spot of the sequence by job $j = 66$.

Table 8.9 MORA results with measures of $B(j)$ for each candidate Job

		k				
j	f	1	2	3	4	B(j)
51		2	1	1	1	−1
52		2	2	1	2	2
53		2	2	0	2	−1
54		2	2	1	1	1
55		2	1	0	2	−3
56		1	2	0	4	−1
57		2	1	1	3	−2
58		3	2	0	3	−1
59		1	3	0	0	−3
60		4	1	0	0	−5
61		1	1	0	3	−4
62		3	2	0	2	1
63		4	2	0	4	−2
64		4	0	1	3	−2
65		0	1	0	1	−3
66		5	2	1	2	3
67		4	1	0	2	−3
68		5	0	1	1	0
69		1	1	0	2	−2
70		5	2	0	4	−1

Table 8.10 Bill-of-material for part, *h*, units per product, *b*, feature, *f*, and element, *e*

h	b	f	e
1	2		3
2	1	3	6
3	4		10
4	2		11
5	1	1	12
6	1		15
7	1	2	16
8	1	4	19

Bill-of-Material

Table 8.10 provides the bill-of-material data for the products of the shift. The list shows $Nh = 8$ parts are needed. The table gives the quantity of each part on every unit of product. Also, when a part is associated with a feature *f*, the feature is noted. Finally, the list gives the work element *e*, that first requires the use of the part. Note, for example, one unit of part $h = 2$ is needed on each product, and the part is associated with feature $f = 3$ and element $e = 6$.

Part Requirements

A final step in the assembly management is to determine the inventory requirements for each of the parts over the shift. Table 8.11 is a worksheet for the calculations. The table gives a list of the parts, *h*. Note, for any element with a feature, the parts may be different for each option. For feature $f = 3$, with only one option, there is only one part, $h = 2$. But for feature $f = 1$, there are five options and thereby, five parts denoted as, 5.1, 5.2, 5.3, 5.4 and 5.5. In the same way, there are three versions of part $h = 7$ and four of $h = 8$. For each of the feature parts, the feature *f*, and option *k*, are listed. The table further identifies the element *e*, that first uses the part. For the parts with a feature, recall, the probability (estimate) *p*, of the part being used on a unit of product is provided. For those parts with no feature, the probability is $p = 1.00$. For the parts with a feature, the probabilities are taken from those provided earlier in Table 8.2. The probabilities *p*, and the shift schedule, *N*, are now combined to compute the estimated frequency of part *h*, over a typical shift, as $N_h = p \times N = p \times 50$. The shift requirement for each part h is computed by $R_h = N_h \times b$, where *b* is the quantity of part units per product, as listed on the bill-of-material. Table 8.4 identifies the stations *i*, where each of the elements are assigned along the line. These are obtained from the line balancing results.

Table 8.11 Worksheet for part h, quantity per product *b*, feature *f*, option *k*, element *e*, probability usage *p*, shift usage N_h, shift requirement R_h, and station *i*

h	b	f	k	e	p	N_h	R_h	i
1	2			3	1.00	50	100	2
2	1	3	1	6	0.50	25	25	1
3	4			10	1.00	50	200	1
4	2			11	1.00	50	100	1
5.1	1	1	1	12	0.20	10	10	2
5.2	1	1	2	12	0.20	10	10	2
5.3	1	1	3	12	0.25	12.5	12.5	2
5.4	1	1	4	12	0.15	7.5	7.5	2
5.5	1	1	5	12	0.10	5	5	2
6	1			15	1.00	50	50	3
7.1	1	2	1	16	0.30	15	15	3
7.2	1	2	2	16	0.40	20	20	3
7.3	1	2	3	16	0.10	5	5	3
8.1	1	4	1	19	0.30	15	15	2
8.2	1	4	2	19	0.30	15	15	2
8.3	1	4	3	19	0.20	10	10	2
8.4	1	4	4	19	0.10	5	5	2

Summary

This chapter summarizes the management steps needed to control a mixed model make-to-order assembly line. The basic data are the jobs on the order board, with due dates awaiting their turn on the line. Every job specifies the combination of options for each feature offered on the products. In this way, each job is a unique item with its own bill-of-material. Over the planning horizon, the projection of features and options is used along with the daily shift schedule quantity to estimate the time needed by work element over the shift. This data along with a precedence diagram is used to assign work elements to the operators on the line where the shift workload is evenly distributed. Each day, a set of jobs is assigned for assembly, and each has its own set of options by feature. A sequence algorithm is used to generate the order the jobs go down the line. This arrangement is made several days before the shift date. Should a job be subsequently removed, prior to the assembly date, an algorithm selects another job from a pool of candidate jobs to insert as a substitute in its place. Finally, the bill-of-material data is called to determine the requirement needs for each part at each station along the line.

Chapter 9
Postponement Assembly

Introduction

Postponement is a strategy that can be applied to products such as trucks, automobiles, farm tractors, and computers that are offered with a variety of features and options. In the assembly process, the units are built without the variety of features and options. The assembly is like a single model line and the output units are stocked in a warehouse facility. When the customer orders come in with the exact feature and option combination, the final assembly takes place in the warehouse. This way, complicated make-to-order assembly is replaced with the simpler single model assembly. This strategy yields less inventory in the plant and reduces the lead time to customers. For convenience in this chapter, the strategy is called full postponement. Two alternative assembly strategies for this environment are demonstrated in comparison: no postponement and partial postponement.

In the inventory operation of these products, the units are produced and stocked in a generic form that only includes the standard components. Subsequently, when the customer orders come in and the specifications are known, a generic unit in stock is then customized with the exact components as requested. In this way, the value of the unit is delayed until the last possible moment and the inventory in stock is reduced accordingly. Further, the postponement strategy reduces the response time from the customer order date to the delivery date. The strategy requires a high degree of cooperation and data transfer across the supply chain. Many manufacturers and retailers are now using postponement to hold inventory in a less finished state in the product assembly until actual customer demand is known. The inventory in the warehouse, in a lean basic form, requires light assembly and packaging before the order is filled. For those products that are adaptable, this supply chain strategy reduces inventory and improves customer service.

Example 9.1 The example data provided is the same for the three assembly strategies described in this chapter. The data is also the same as given in Chap. 8, Mixed Model Make-to-Order Assembly. The shift time is $T = 450$ min, the products are associated with features, f, and options, k.

N. T. Thomopoulos, *Assembly Line Planning and Control*,
DOI: 10.1007/978-3-319-01399-2_9,

The example data is described in the sections called: work elements, features and options and bill-of-material, and is listed in Tables 9.1, 9.2 and 9.3, respectively. As in all the chapters of this book, the example data is small compared to a typical assembly entity, but is sufficient enough to allow the reader to follow the methodologies. Altogether there are 28 elements, four features, various options and eight parts.

Work Elements

When considering the postponement strategy for a line, the data the assembly management gathers is essentially the same as a make-to-order line. The work elements are identified with related data as listed in Table 9.1. The table contains the elements, e, the element time, t_e, the predecessor elements, p, and the associated features, f, if any.

Table 9.1 Work elements, e, element times, t_e, predecessor elements, p, and features, f

e	t_e	p	f
1	0.8	6	
2	2.8	3	
3	1.6		
4	1.2	6	
5	1.1	6	
6	2.5		3
7	0.5	1	
8	0.7	7	
9	0.7	5	
10	0.6	6	
11	0.8	10	
12	2.4	19	1
13	1.2	11	
14	1.9	15	
15	0.7	8	
16	1.9	12	2
17	0.8	13	
18	1.5	17	
19	2.2	2	4
20	0.4	12	
21	0.9	17	
22	1.4	14, 16	
23	1.2	21	
24	1.3	23	
25	0.9	23	
26	2.7	14	
27	1.6	26	
28	1.3	27	

Table 9.2 Probabilities, $P(f, k)$, of option, k, by feature, f

$f\backslash k$	0	1	2	3	4	5
1	0.10	0.20	0.20	0.25	0.15	0.10
2	0.20	0.30	0.40	0.10		
3	0.50	0.50				
4	0.10	0.30	0.30	0.20	0.10	

Table 9.3 Bill-of-material for part, h, units per product, b, feature, f, and element, e

h	b	f	e
1	2		3
2	1	3	6
3	4		10
4	2		11
5	1	1	12
6	1		15
7	1	2	16
8	1	4	19

Features and Options

Another basic data need is the estimate of the probability of options by feature, $P(f,k)$, as listed in Table 9.2. The table shows four features, f, and the list of options for each feature. Recall, option $k = 0$ is a null option where the customer does not want any option of feature f. The sum of the probabilities over a feature is one.

Bill-of-Material

Another set of data associated with the assembly line is the bill-of-material information as reported in Table 9.3. The table gives a list of the parts, h, that are needed on the mix of products, the bom quantity, b, of each part per unit of product, the feature, f, associated with the part, and the element, e, that first uses the part.

No Postponement (Make-to-Order Assembly)

The products of this chapter have features and options and the assembly will be a make-to-order type. The units are individually produced to the specifications of the customers. Each unit comes with a list of features and every feature has a variety of options. The customer selects the options wanted for every feature, and sometimes the customer bypasses a particular feature altogether, called the null option. The combination of options and features make each unit a unique model. Often, for an assembly shift, no two units going down the line are the same. The units are called jobs and the job has a list of specifications by feature and option. Every job also has a due date as promised to the customer.

When no postponement takes place, the jobs are assigned to a shift prior to their assembly down the line when all the parts and components needed for the job are available and as close to the job due date as possible. After all the jobs are selected for a shift, the sequence of the jobs is decided and this takes place several days prior to assembly. The sequencing seeks to spread the jobs down the line in a manner where the options by feature are as far apart as possible. This is needed since some of the options on a feature take longer to process than others. An algorithm to accomplish the sequence is provided in Chap. 8. With the sequence available several days in advance of the shift, the management can arrange to have all the particular parts and components by job ready at the assigned stations along the line, and in that way the assembly process can take place in a smoothed manner.

The example here is taken from Chap. 8 that describes make-or-order assembly. A quick review is provided here on running the products using the assembly system, and the results are the same result as in Chap. 8. Recall the shift time is $T = 450$ min, and the shift schedule calls for $N = 50$ units.

Shift Element Times

Table 9.4 is a worksheet that determines the shift time per element, T_e. The table lists the frequency, N_e, of use for each element over the shift. For those elements that are not associated with a feature, the frequency is $N_e = N = 50$. For those elements with a feature connection, $N_e = N \times [1 - p(f,0)]$, where $p(f,0)$ is the probability of the null option for the feature as listed in Table 9.2. So for element $e = 1$ that is not connected with a feature, $N_e = 50$. Element $e = 3$, connected to feature $f = 3$, has $N_e = N \times [1 - p(3,0)] = 50 \times [1 - 0.50] = 25$. The shift times for each of the elements become $T_e = N_e \times t_e$, where t_e is the element time.

Line Balancing

The sum of shift times per element is $\Sigma T_e = 1776$ min. With this information, the number of stations needed on the line becomes $n = 4$. The average shift time per station becomes, $\bar{T} = \sum T_e/n = 1776/4 = 444$ minutes. Table 9.5 gives the line balancing results for each of the four stations, i. Recall, the goal is to assign the elements, e, to the stations in a way where the station shift times, T_i, are close to

Table 9.4 Elements, e, element time, t_e, features, f, shift element frequency, N_e, and shift element time, T_e

e	t_e	f	N_e	T_e
1	0.8		50	40
2	2.8		50	140
3	1.6		50	80
4	1.2		50	60
5	1.1		50	55
6	2.5	3	25	62.5
7	0.5		50	25
8	0.7		50	35
9	0.7		50	35
10	0.6		50	30
11	0.8		50	40
12	2.4	1	45	108
13	1.2		50	60
14	1.9		50	95
15	0.7		50	35
16	1.9	2	40	76
17	0.8		50	40
18	1.5		50	75
19	2.2	4	45	99
20	0.4		50	20
21	0.9		50	45
22	1.4		50	70
23	1.2		50	60
24	1.3		50	65
25	0.9		50	45
26	2.7		50	135
27	1.6		50	80
28	1.3		50	65
Sum				1776

$\bar{T} = 444$ minutes and all of the predecessor element constraints listed in Table 9.1 are satisfied. The table lists the stations, i, elements, e, element times, t_e, features, f, shift element times, T_e, and shift station times, T_i. Note the shift station times become: 443, 447, 441, and 445 min for stations, 1–4, respectively. Also note, the sum of the shift station times is 1776 min.

Line Sequencing

Each day, the line sequencing of the jobs takes place. The sequence algorithm places the jobs in order down the line in a way where the options of each of the features are spaced as far apart as possible. For brevity, the sequencing logic is not repeated. The reader can refer back to Chap. 8 for the full sequencing detail.

Table 9.5 Station i, element, e, element time, t_e, feature, f, shift element time, T_e, and shift station time, T_i

i	e	t_e	f	T_e	T_i
1	1	0.8		40	
	4	1.2		60	
	5	1.1		55	
	6	2.5	3	62.5	
	7	0.5		25	
	8	0.7		35	
	9	0.7		35	
	10	0.6		30	
	11	0.8		40	
	13	1.2		60	443
2	2	2.8		140	
	3	1.6		80	
	12	2.4	1	108	
	19	2.2	4	99	
	20	0.4		20	447
3	15	0.7		35	
	16	1.9	2	76	
	17	0.8		40	
	18	1.5		75	
	21	0.9		45	
	23	1.2		60	
	24	1.3		65	
	25	0.9		45	441
4	14	1.9		95	
	22	1.4		70	
	26	2.7		135	
	27	1.6		80	
	28	1.3		65	445
Sum					1776

Shift Part Requirements

The part requirement for a shift is shown in Table 9.6. This is a worksheet for a typical day. The data in this table comes from Tables 9.1, 9.2, 9.3, and 9.5. The table lists the parts, h, bom quantity of parts, b, per unit of product, feature, f, option, k, associated element, e, probability of usage, p, element frequency in a shift duration, N_e, part shift requirement, R_h, and assigned station, i. Note some parts with a feature connection, may have multiple variants. Part $h = 2$ with feature $f = 3$ has only one variant, since the feature has only one non-null option. But part $h = 5$, associated with feature $f = 1$ that has five options, thereby has five part variants, 5.1, 5.2, 5.3, 5.4, 5.5. For the parts without a feature connection, the probability usage, p, is 1.00. For the parts with a feature connection, the probability usage, p, comes from the list in Table 9.2. The element shift usage is

$N_e = p \times N = p \times 50$. The part shift requirement becomes $R_h = N_e \times b$. Finally, the first station that needs the part, i, is listed. Note, $e = e(h)$ in the table.

Full Postponement (Single Model Assembly)

The example continues where now the management explores the use of full postponement. The strategy calls for not inserting any of the features that are of the non-standard type. In the example, four features with a variety of options are identified. The assembly will not include these features and their associated options. So, the work elements and parts that are connected to the features are not included in the assembly processing. In essence, all the units going down the line have only the standard parts and components, and thereby are all the same. The units are produced as a basic lean single model. This way, the units produced are the same as single model assembly. The task of assigning the work elements to the operators in a balanced way without violating any precedence constraints is still needed. This is the role of line balancing as in single model assembly. Now since, the units of assembly are all of the same type, no sequencing algorithm is needed. The units just go down the line one after the other.

Recall, when the units are completed in assembly, they are stocked in the warehouse facility in a lean form without any feature and options. As each customer order arrives, and the exact specification of features and options are known, the final assembly takes place in the warehouse accordingly.

Table 9.6 Worksheet for part h, bom quantity per product, b, feature, f, option, k, element, e, probability usage, p, shift usage, N_e, shift requirement, R_h, and station, i

h	b	f	k	e	p	N_e	R_h	i
1	2			3	1.00	50	100	2
2	1	3	1	6	0.50	25	25	1
3	4			10	1.00	50	200	1
4	2			11	1.00	50	100	1
5.1	1	1	1	12	0.20	10	10	2
5.2	1	1	2	12	0.20	10	10	2
5.3	1	1	3	12	0.25	12.5	12.5	2
5.4	1	1	4	12	0.15	7.5	7.5	2
5.5	1	1	5	12	0.10	5	5	2
6	1			15	1.00	50	50	3
7.1	1	2	1	16	0.30	15	15	3
7.2	1	2	2	16	0.40	20	20	3
7.3	1	2	3	16	0.10	5	5	3
8.1	1	4	1	19	0.30	15	15	2
8.2	1	4	2	19	0.30	15	15	2
8.3	1	4	3	19	0.20	10	10	2
8.4	1	4	4	19	0.10	5	5	2

Work Elements

Table 9.7 is a list of the work element data for the full postponement example. The table contains the elements, e, element times, t_e, predecessor elements, p and the features. Because, a full postponement is in place, the elements with the features are voided. These are element 6, 12, 16, and 19. The element times for these are set to zero. The predecessor elements are still listed, but since the element times are zero, they have no affect in assigning the elements to stations. The sum of the element times becomes $\Sigma t_e = 28.6$ min.

Note that sometimes another element, not the prime element, will also not be used since it applies only to the feature that is dropped for postponement. The example here does not show any of these elements.

Shift Schedule

The element times sum to $\Sigma t_e = 28.6$ min, and the shift time is $T = 450$ min. Assuming $n = 4$ stations will still be in place, the number of units to assemble over the shift is computed as below:

$$N' = n \times T \Big/ \sum t_e = 4 \times 450/28.6 = 62.94$$

Rounding down to an integer, gives,
$N = 62$ units.

Line Balancing

A next step is to assign the elements to the four stations where the station cycle time, c_i, $i = 1$ to 4 are relatively the same. With $\sum t_e = 28.6$ and $n = 4$, the average station time becomes,

$$\overline{c} = \sum t_e/n = 28.6/4 = 7.15\,\text{min}$$

So as much as possible, the elements are assigned to the four stations in a way where the station times are close to 7.15 min, and all of the precedence restrictions are in compliance. The line balancing results are listed in Table 9.8. The table lists the stations, i, elements, e, element times, t_e, and station times, c_i. The station times are: 7.3, 6.6, 7.7, and 7.0 min. The cycle time for the line becomes

$$c = \max(7.3, 6.6, 7.7, 7.0) = 7.7\,\text{min}.$$

Table 9.7 Work elements, e, element times, t_e, predecessor elements, p, and features, f

e	t_e	p	f
1	0.8	6	
2	2.8	3	
3	1.6		
4	1.2	6	
5	1.1	6	
6	0.0		3
7	0.5	1	
8	0.7	7	
9	0.7	5	
10	0.6	6	
11	0.8	10	
12	0.0	19	1
13	1.2	11	
14	1.9	15	
15	0.7	8	
16	0.0	12	2
17	0.8	13	
18	1.5	17	
19	0.0	2	4
20	0.4	12	
21	0.9	17	
22	1.4	14, 16	
23	1.2	21	
24	1.3	23	
25	0.9	23	
26	2.7	14	
27	1.6	26	
28	1.3	27	
Sum	28.6		

Thereby, the effective number of units that will be completed in a shift of $T = 450$ min is,

$$N = T/c = 450/7.7 = 58.44$$

that rounds to $N - 58$.

The efficiency of the line is measured below:

$$E = \bar{c}/c = 7.15/7.7 = 0.929.$$

Table 9.8 Station, i, element, e, element time, t_e, and station time, c_i

i	e	t_e	c_i
1	1	0.8	
	4	1.2	
	6	0	
	7	0.5	
	8	0.7	
	10	0.6	
	11	0.8	
	13	1.2	
	15	0.7	
	17	0.8	7.3
2	5	1.1	
	9	0.7	
	2	2.8	
	3	1.6	
	12	0	
	16	0	
	19	0	
	20	0.4	6.6
3	14	1.9	
	18	1.5	
	21	0.9	
	23	1.2	
	24	1.3	
	25	0.9	7.7
4	22	1.4	
	26	2.7	
	27	1.6	
	28	1.3	7.0

Shift Part Requirements

Table 9.9 is the worksheet for the shift part requirements on this full postponement example. The only parts that have requirements are those with no connection to the features. Table 9.9 is based on the line balance results that show the effective shift schedule will be 58 units. Note the drop in the total inventory needs when postponement is run instead of make-to-order assembly (Table 9.6).

Partial Postponement (Make-to-Stock Assembly)

Assume now, where the management considers a partial postponement option. In the example, only feature $f = 2$ is applied in the assembly, and thus, this is a partial postponement situation. Recall from Table 9.2, the option probabilities for

Table 9.9 Part, h, usage bom quantity, b, feature, f, option, k, element, e, shift usage, N_e, shift part requirement, R_h, and station, i

h	b	f	k	e	N_e	R_h	i
1	2			3	58	116	2
2	1	3	1	6	0	0	0
3	4			10	58	232	1
4	2			11	58	116	1
5.1	1	1	1	12	0	0	0
5.2	1	1	2	12	0	0	0
5.3	1	1	3	12	0	0	0
5.4	1	1	4	12	0	0	0
5.5	1	1	5	12	0	0	0
6	1			15	58	58	1
7.1	1	2	1	16	0	0	0
7.2	1	2	2	16	0	0	0
7.3	1	2	3	16	0	0	0
8.1	1	4	1	19	0	0	0
8.2	1	4	2	19	0	0	0
8.3	1	4	3	19	0	0	0
8.4	1	4	4	19	0	0	0

feature $f = 2$ are: 0.2, 0.3, 0.4, and 0.1, respectively, for options $k = 0, 1, 2$, and 3. In this partial postponement assembly, the units to assemble will be either of four types, in accordance with the probabilities cited. 20 % will have none of the four features, 30 % will have option 1 of feature 2, 40 % will have option 2 of feature 2, and 10 % will have option 3 of feature 2. The element associated with feature $f = 2$ is $e = 16$. Assume from the time studies, the standard times for the four options are: 0.0, 2.3, 1.5, and 2.3 min for options 0–3, respectively. Essentially, the assembly system will be processing four different type of units, and therefore, these are here called four models, (0, 1, 2, 3). The weighted average time for the element $e = 16$ is computed as below:

$$t_e = 0.2 \times 0.0 + 0.3 \times 2.3 + 0.4 \times 1.5 + 0.1 \times 2.3 = 1.52\,\text{min}$$

This partial postponement strategy is, in effect, run as a mixed model make-to-stock assembly line, where the models are identified by the four options of feature f − 2.

The management of this line still must assign the work elements to the operators in a way where the shift station times are fairly equal and where the element precedence constraints are not violated. This is the line balancing method as shown in Chap. 7, Mixed Model Make-to-Stock Assembly. Further, since a finite number of models are being processed on the line, the management could use the sequencing algorithm (MSSA) described in the chapter. Typically, the line balancing assignments are not changed for all the days over the planning horizon. Should the daily schedules have slight variations in the model mix, a new sequence of models would need to be generated each day as well.

Table 9.10 Work elements, e, element times, t_e, predecessor elements, p, and features, f

e	t_e	p	f
1	0.8	6	
2	2.8	3	
3	1.6		
4	1.2	6	
5	1.1	6	
6	0.0		3
7	0.5	1	
8	0.7	7	
9	0.7	5	
10	0.6	6	
11	0.8	10	
12	0.0	19	1
13	1.2	11	
14	1.9	15	
15	0.7	8	
16	1.52	12	2
17	0.8	13	
18	1.5	17	
19	0.0	2	4
20	0.4	12	
21	0.9	17	
22	1.4	14,16	
23	1.2	21	
24	1.3	23	
25	0.9	23	
26	2.7	14	
27	1.6	26	
28	1.3	27	
Sum	30.12		

Work Elements

The work elements are listed in Table 9.10, along with the element times, t_e, predecessors, p, and associated features, f. The elements without a feature connection have no change in element times. The elements with a feature that is not the selected feature ($f = 1, 3, 4$) have zero element times and thereby are not used in processing the units. The element ($e = 16$) that is associated with feature $f = 2$ does have a change in the element time. The element time listed for $e = 16$ ($t_e = 1.52$ min) is a weighted average of the element times for each of the options from feature $f = 2$. Recall, for element $e = 16$, $t_e = 1.52/0.8 = 1.9$ is the weighted time for all but the null option. Note, the sum of the element times ($\Sigma t_e = 30.12$ min) is listed at the bottom of the table.

Shift Assembly Schedule

The management can now determine the number of units to process in a shift. The shift schedule quantity, N, is computed using the following data, $T = 450$ min = shift time, $n = 4$ = number of stations, and $\Sigma t_e = 30.12$ min = weighted average unit time. So now,

$$N = n \times T/\Sigma t_e = (4 \times 450)/30.12 = 59.76\,\text{units}$$

that rounds down to $N = 59$.

Shift Element Times

Since the number of stations remains at $n = 4$, and the shift schedule is set at $N = 59$, the element shift times, T_e, can now be computed. These are by $T_e = N \times t_e$, where t_e is the work time for element e. Recall, for element $e = 16$, the element time is $t_e = 1.52$ min, representing the weighted average time for all four options (or models). In this way, the frequency of shift usage per element e is $N_e = 59$. The shift element times, T_e, are computed in Table 9.11 that serves as a worksheet. The table lists the elements, e, the element times, t_e, the element usage, N_e, and the element shift time, T_e. The shift usage is $N_e = 59$. The bottom of the table shows the sum of times for the unit, $\Sigma t_e = 30.12$ min, and the total shift time is $\Sigma T_e = 1777.1$ min.

Line Balancing

Assuming $n = 4$ stations, the line balancing can be performed. The average station time will be $\overline{T} = \Sigma T_e/n = 1777.1/4 = 444.3$ minutes. The goal is to assign the elements to the stations with station times near 444.3 min, and with all precedence constraints followed. One solution is shown in Table 9.12 where the station shift times are 448.4, 437.8, 454.3, and 436.6 min, respectively for stations 1 to 4.

The efficiency of the line can now be measured. This is as follows,

$$E = \overline{T}/T = 444.3/454.3 = 0.978$$

or 97.8 %.

Table 9.11 Element e, element time t_e, element shift frequency N_e, and element shift time T_e

e	t_e	N_e	T_e
1	0.8	59	47.2
2	2.8	59	165.2
3	1.6	59	94.4
4	1.2	59	70.8
5	1.1	59	64.9
6	0	0	0
7	0.5	59	29.5
8	0.7	59	41.3
9	0.7	59	41.3
10	0.6	59	35.4
11	0.8	59	47.2
12	0	0	0
13	1.2	59	70.8
14	1.9	59	112.1
15	0.7	59	41.3
16	1.52	59	89.7
17	0.8	59	47.2
18	1.5	59	88.5
19	0	0	0
20	0.4	59	23.6
21	0.9	59	53.1
22	1.4	59	82.6
23	1.2	59	70.8
24	1.3	59	76.7
25	0.9	59	53.1
26	2.7	59	159.3
27	1.6	59	94.4
28	1.3	59	76.7
Sum	30.12		1777.1

Line Sequencing

Recall how the partial postponement forms a finite variation in the units of product. The units become essentially different models. Each day, a slight difference in the mix of models may be called, and thereby a new sequence for the day is needed. The sequencing algorithm of Chap. 7 (MSSA) spaces the models as far apart and can be used for this application. For brevity, the description of the sequencing algorithm is not repeated here.

Table 9.12 Station i, element e, element time t_e, element shift frequency N_e, element shift time T_e and station shift time T_i

i	e	t_e	N_e	T_e	T_i
1	1	0.8	59	47.2	
	4	1.2	59	70.8	
	6	0	0	0	
	7	0.5	59	29.5	
	8	0.7	59	41.3	
	10	0.6	59	35.4	
	11	0.8	59	47.2	
	13	1.2	59	70.8	
	15	0.7	59	41.3	
	5	1.1	59	64.9	448.4
2	9	0.7	59	41.3	
	2	2.8	59	165.2	
	3	1.6	59	94.4	
	12	0	0	0	
	16	1.52	59	89.7	
	19	0	0	0	
	17	0.8	59	47.2	437.8
3	14	1.9	59	112.1	
	18	1.5	59	88.5	
	21	0.9	59	53.1	
	23	1.2	59	70.8	
	24	1.3	59	76.7	
	25	0.9	59	53.1	454.3
4	20	0.4	59	23.6	
	22	1.4	59	82.6	
	26	2.7	59	159.3	
	27	1.6	59	94.4	
	28	1.3	59	76.7	436.6
Sum		30.12		1777.1	1777.1

Shift Part Requirements

Table 9.13 is the worksheet for the shift part requirements on this partial postponement example. The results are based on the shift requirements of $N - 59$ units. The only parts that have requirements are those with no connection to the features, except for feature $f = 2$, and these are denoted as: 7.1, 7.2, and 7.3. Note, for example, the shift frequency for part $h = 7.1$ is $Ne = 0.3 \times 59 = 17.7$, and this rounds to $N_e = 18$. The parts connected to features $f = 1, 3, 4$ do not have any inventory requirements. The table lists the stations where the inventory is needed, as per the line balance results. Note, the difference in inventory needs from the make-to-order and the full postponement examples.

Table 9.13 Part *h*, bom quantity *b*, feature *f*, option *k*, element *e*, element frequency N_e, part requirement, R_h, and station *i*

h	b	f	k	e	N_e	R_h	i
1	2			3	59	118	2
2	1	3	1	6	0	0	0
3	4			10	59	236	1
4	2			11	59	118	1
5.1	1	1	1	12	0	0	0
5.2	1	1	2	12	0	0	0
5.3	1	1	3	12	0	0	0
5.4	1	1	4	12	0	0	0
5.5	1	1	5	12	0	0	0
6	1			15	59	59	1
7.1	1	2	1	16	18	18	2
7.2	1	2	2	16	23	23	2
7.3	1	2	3	16	6	6	2
8.1	1	4	1	19	0	0	0
8.2	1	4	2	19	0	0	0
8.3	1	4	3	19	0	0	0
8.4	1	4	4	19	0	0	0

Summary

Postponement is a supply chain management strategy to reduce the inventory needs of the parts and components, and also lower the lead time to customers. The strategy applies for make-to-order mixed model assembly lines that have a series of features and options. The units are assembled in a generic way without any of the options, and are stored in a warehouse awaiting the customer orders. As the customer orders arrive with the exact options specified, the final assembly takes place. This strategy of full postponement is compared to two other strategies of no postponement and of partial postponement. The chapter shows how to assign the elements to the operators, and how to sequence the units down the line. Also, the bill-of-material data is applied to determine the requirement needs for the shift and for each station.

Chapter 10
One Station Assembly

Introduction

One station assembly is described in the context of a shoe manufacturing plant where one worker is assigned a set of shoes by style and size to assemble all alone. The worker is given a batch of the items to produce, and is provided all the parts and components needed to complete the task. Multiple pairs are assigned to the worker; typically six to twelve pair at one time. The worker completes all the pairs in the batch prior to starting the next assignment. This is an example of one-station assembly. In other situations, the workers are assigned one unit at a time, as in engine assembly. This chapter describes some of the quantitative methods that are related to one station assembly. Sometimes the station operator requires a mold of some type to carry out his/her work. The mold is used in the production process and then can subsequently be used for another unit. The plant has an inventory of molds to allow the workers to carry out their assignments. This chapter shows how to determine the number of molds to have in the plant in order to yield a specified service level. The service level (SL) is the probability the mold will be available when needed by a worker.

Inventory and Requirement Data by Model

Example 10.1 Suppose a plant produces five models, $Nj = 5$, in a make-to-stock manner. The plant schedule calls for a steady flow of the models each day to achieve an efficient production process. Every day the plant schedules a fixed number of units for production. The way to determine the models to schedule for the day begins with the inventory status of each of the models as described below.

Consider the stock status data for five models as listed in Table 10.1. The models are denoted as $j = 1$–5. The table shows the current levels of on-hand, oh, and on-order, oo, inventory. In addition, the table lists the daily requirements, r, for the planning horizon of 5 days. For simplicity, the planning covers only the

N. T. Thomopoulos, *Assembly Line Planning and Control*,
DOI: 10.1007/978-3-319-01399-2_10,

Table 10.1 On-hand, oh, on-order, oo and five daily requirements *r*, by model, *j*

			--r--				
j	oh	oo	1	2	3	4	5
1	37	3	10	10	10	10	10
2	24	0	20	20	20	20	20
3	41	9	30	30	30	30	30
4	58	12	40	40	40	40	40
5	73	0	50	50	50	50	50
Sum	233	24	150	150	150	150	150

5 future days. Also for simplicity, the daily requirements are all the same, while in actual practice they often vary from day to day. The bottom of the table gives the corresponding sum quantities for all the models. This is the primary data needed to determine the plant schedule for the coming days over the planning horizon. Below shows a method to compute the plant schedule that seeks an equal level of days-supply for each of the models.

Days-Supply by Model

The term days-supply (ds), represents the number of days from the future requirement that is available in the current inventory, oh-hand plus on-order. This quantity is measured for each of the models, and the results are listed in Table 10.2. The Table notes the models, *j*, the sum of on-hand plus on-order inventory, ohoo, the average daily requirements, *r*, and the ds. Note for model, $j = 1$, since ohoo $= 40$ and $r = 10$, the ohoo will cover the requirements for 4 future days, thus, ds $= 4.00$. In the same way, the ds for all of the models are listed. The smaller the days-supply measure, ds, the sooner the model needs new inventory.

Days-Supply for All Models

Assume the daily plant schedule calls for Q units of product. This is the quantity of new stock inventory. In the example, suppose $Q = 60$ units. With this quantity, and with the sum of on-hand and on-order inventory, $\sum$(ohoo)$=\sum$(oh + oo),

Table 10.2 On-hand plus on-order, ohoo, average daily requirements, *r*, and days-supply ds, by model, *j*

j	ohoo	*r*	ds
1	40	10	4.00
2	24	20	1.20
3	50	30	1.67
4	70	40	1.75
5	73	50	1.46
Sum	257	150	

shown in Table 10.1, the ds for the aggregate of all models is now computed as below:

$$\begin{aligned} \text{ds} &= \left[\sum(\text{ohoo}) + Q\right] / \sum(r) \\ &= [257 + 60]/150 = 2.11 \end{aligned}$$

This measure shows that the current inventory plus the schedule quantity is adequate to cover the coming 2.11 days of future requirements. When, comparing this measure with the ds measure of Table 10.2, model $j = 1$ with ds $= 4.00$ is high (larger than 2.11) in stock and the other models are low. So, the aggregate ds will be adjusted by not using the data from model $j = 1$.

Adjusted Days-Supply by Model

Table 10.3 shows how the ds is computed without the data from model $j = 1$. The sum of the on-hand and on-order is now $\sum(\text{ohoo}) = 217$and the corresponding sum of the requirements is $\sum(r) = 140$. Using the plant schedule of $Q = 60$, the aggregate ds becomes:

$$\text{ds} = [217 + 60]/140 = 1.9785 \cong 1.98$$

Note, the model measures of ds now are all lower than 1.98 days.

Build Quantity by Model

The goal is to develop a plant schedule by model so the ds by model is as even as possible. A worksheet to do this is shown in Table 10.4. The table pertains only to the models that require new stock ($j = 2, 3, 4, 5$), and lists the on-hand plus on-order, ohoo, the average daily requirements, r, and the days-supply, ds. For each of the models, the model requirements for the aggregate days-supply (ds $= 1.9785$) are computed and designated as $r(1.98)$. Note for model $j = 2$,

Table 10.3 Adjusted on-hand plus on-order, ohoo, average daily requirements, r, and days-supply ds, by model, j

j	ohoo	r	ds
1	-	-	-
2	24	20	1.20
3	50	30	1.67
4	70	40	1.75
5	73	50	1.46
Sum	217	140	

Table 10.4 On-hand plus on-order, ohoo, average daily requirements, r, days-supply ds, 1.98 days requirement $r(1.98)$, raw quantity, q', with ds$'$, and rounded quantity, q_o, with ds$_o$ by model j

j	ohoo	r	ds	$r(1.98)$	q'	ds$'$	q_o	ds$_o$
1	-	-						
2	24	20	1.20	39.57	15.57	1.98	15	1.95
3	50	30	1.67	59.35	9.35	1.98	9	1.97
4	70	40	1.75	79.14	9.14	1.98	9	1.97
5	73	50	1.46	98.94	25.94	1.98	27	2.00
Sum	217	140		277.00	60.00		60	

where $r(1.98) = 1.9785 \times r = 1.9785 \times 20 = 39.57$. The quantity of new stock to build for model $j = 2$ would then be,

$$\begin{aligned} q' &= r(1.98) - \text{ohoo} \\ &= 39.57 - 24 = 15.57 \end{aligned}$$

In the same way, the requirements, $r(1.98)$ and the build quantities, q', for the remaining models are computed. Note, the sum of the build quantities is $\sum[q'] = 60$, and the ds for models $j = 2$, 3, 4 and 5 become ds$' = 1.98$.

But, the build quantities for the models must be in integers, and not fractions. Further, in many plants, the quantities must also be set in multiple quantities, M. Assume in this example, the multiple quantities are set as $M = 3$ units. With this restriction, the model build quantities, now denoted as q_o, are reset in a way where all are in multiples of $M = 3$ and where the sum is still $Q = \sum(q_0) = 60$. The computations show where the build schedule for the day becomes: $q_o = (0, 15, 9, 9, 27)$ for $j = 1$, 2, 3,4 5, respectively.

The three measures of ds in the table are calculated using the following:

$$\begin{array}{lll} \text{ds} & \text{uses} & \text{ohoo,} \\ \text{ds}' & \text{uses} & (\text{ohoo} + q') \\ \text{ds}_o & \text{uses} & (\text{ohoo} + q_o) \end{array}$$

The quantity q' will yield the exact ds (1.98) to each model, and the quantity q_o is rounded to comply with the multiple of $M = 3$.

Build Schedule of Model at Station

The example continues and assumes the production plant will distribute the work to five operators, denoted as $i = 1, 2, 3, 4, 5$. The management now allocates the build quantity of $Q = 60$ units to each of the five operators with $Q/5 = 12$ units each. In the example, assume the units are also restricted to multiples of $M = 3$. One such allotment of the schedule is shown in Table 10.5. Operator $i = 1$, is

Table 10.5 Build schedule N_{ij}, for station i, on model j

	i				
j	1	2	3	4	5
1	0	0	0	0	0
2	6	6	3	0	0
3	0	0	3	6	0
4	3	0	0	0	6
5	3	6	6	6	6

assigned 6 units of $j = 2$, 3 of j = 4 and 3 of $j = 5$. For notation sake, N_{ij} will represent the daily schedule at station i for model j. In the same way, the five operators are assigned their workload in a way where the sum per operator is 12 units and the aggregate sums by model are consistent with the schedule mix ($q = 0, 15, 9, 9, 27$) listed in Table 10.4.

Bill-of-Material

The example continues with the bill-of-material data for each of the models. Suppose eight parts, h, are needed in the assembly process, and these are listed in Table 10.6 by model, j. Notice where $j = 1$ requires one unit of parts $h = 1$ and 2 and two units of part $h = 4$ in the production stage, and so forth.

Part Requirements by Station

Combining the build schedule mix of models by stations and the bill-of-material data, the part h requirements, R_{hij}, at station i and for model j are listed in Table 10.7. The table serves as a worksheet showing the stations, i, models, j, and parts, h. The part requirement sums, by station, are also listed in the table.

Part Requirement at Station

Table 10.8 summarizes the part requirements by station. These are the results obtained in Table 10.7. This represents the inventory needed at the start of the day for each station. The sum of the part requirements is also listed in the table. For part $h = 1$, the requirement is $R_h = 60$ units. In the same way, the part requirements for the day are listed for $h = 1$–8.

Table 10.6 Bill-of-material for part, h, and model j

j	h 1	2	3	4	5	6	7	8
1	1	1	0	2	0	0	0	0
2	1	1	0	0	2	0	0	2
3	1	1	0	0	0	2	0	2
4	1	0	1	0	0	0	2	0
5	1	0	1	0	0	0	0	2

Table 10.7 Part requirements, R_{hij}, at station i for part, h and model j

i	j	N_{ij}	h 1	2	3	4	5	6	7	8
1	1	0	0	0	0	0	0	0	0	0
	2	6	6	6	0	0	12	0	0	12
	3	0	0	0	0	0	0	0	0	0
	4	3	3	0	3	0	0	0	6	0
	5	3	3	0	3	0	0	0	0	6
sum			12	6	6	0	12	0	6	18
2	1	0	0	0	0	0	0	0	0	0
	2	6	6	6	0	0	12	0	0	12
	3	0	0	0	0	0	0	0	0	0
	4	0	0	0	0	0	0	0	0	0
	5	6	6	0	6	0	0	0	0	12
sum			12	6	6	0	12	0	0	24
3	1	0	0	0	0	0	0	0	0	0
	2	3	3	3	0	0	6	0	0	6
	3	3	3	3	0	0	0	6	0	6
	4	0	0	0	0	0	0	0	0	0
	5	6	6	0	6	0	0	0	0	12
sum			12	6	6	0	6	6	0	24
4	1	0	0	0	0	0	0	0	0	0
	2	0	0	0	0	0	0	0	0	0
	3	6	6	6	0	0	0	12	0	12
	4	0	0	0	0	0	0	0	0	0
	5	6	6	0	6	0	0	0	0	12
sum			12	6	6	0	0	12	0	24
5	1	0	0	0	0	0	0	0	0	0
	2	0	0	0	0	0	0	0	0	0
	3	0	0	0	0	0	0	0	0	0
	4	6	6	0	6	0	0	0	12	0
	5	6	6	0	6	0	0	0	0	12
sum			12	0	12	0	0	0	12	12

Table 10.8 Part, h, requirements, R_{hi}, at station i

	h							
i	1	2	3	4	5	6	7	8
1	12	6	6	0	12	0	6	18
2	12	6	6	0	12	0	0	24
3	12	6	6	0	6	6	0	24
4	12	6	6	0	0	12	0	24
5	12	0	12	0	0	0	12	12
All	60	24	36	0	30	18	18	102

Plant Reusable Mold Inventory

Often in the production process on one station assembly, the operator requires a mold of some type to build the unit of product. In shoe manufacture, the mold is called a last. The last is needed to fit the leather and all in the shape of the style shoe (model), for the exact size and width. One last is for the right shoe and another for the left shoe. The production of a style shoe with a given size and width cannot take place unless the pair of lasts are available in the plant inventory. This inventory is expensive and takes up much storage space. The last is used to produce a pair of shoes and when done, the last is again usable for another pair of shoes of the same style and size. Thus it is reusable inventory needed to carry out the production in the plant. This is a reusable inventory example in the shoe industry, but molds of some type are needed in building many other products as well, like the molds of the auto windows by year and model of car.

The discussion below gives a way to determine how much of the mold inventory to have in the plant to allow the production process to run smoothly. As mentioned above, the inventory is expensive and is vital. It is important for the plant management to have a systematic way to determine how much of each mold (lasts) to have by model. The method is described below and uses the mathematics of queuing theory.

Queuing Computations

Consider a queuing system with k molds where the customer inter-arrival times and assembly process times are exponentially distributed. When all the molds are in use, the new demands will wait in a queue until one becomes available. The average time between customer demands is denoted as $\tau_a = 1/\lambda$, and the average service time is $\tau_s = 1/\mu$. The following notation applies here:

$\tau_a = 1/\lambda =$ average time between demands
$\tau_s = 1/\mu =$ average time to process a unit
$\lambda =$ average number of demands per unit of time

μ = average number of units processed in a unit of time for a continuously busy service facility
k = number of molds
$\rho = \tau_s/\tau_a = \lambda/\mu$ = utilization ratio
$\rho/k < 1$ is needed to ensure the system is in equilibrium
n = number of units in the system (being process and in queue) where $n \geq 0$
P_n = probability on n units in the system $n = 0, 1, 2, \ldots$
The probability the system is empty ($n = 0$), is listed below.

$$P_0 = 1/\left\{\sum_{n=0}^{k-1} \rho^n/n! + \rho^k/[(k-1)!(k-\rho)]\right\}$$

The probability of n units in the system becomes

$$P_n = \begin{cases} \rho^n/n!P_0 & n = 0 \text{ to } k-1 \\ \rho^n/\left[k!k^{n-k}\right]P_0 & n \geq k \end{cases}$$

The SL is the probability a new demand does not wait for service. This is the probability that n is less than k, $P_{n\,<\,k.}$ Hence,

$$\text{SL} = P_{n\,<\,k}$$

The Table 10.9 gives the minimum number of molds (k) needed to achieve the service level (SL = 0.85, 0.90, 0.95, 0.99) in an infinite queue capacity system with selected values of the utilization ratio (ρ) ranging from 0.1 to 700.

Example 10.2. Consider the shoe manufacturer using a mold (called a 'last') to produce a certain style shoe. The forecast calls for 1,200 pair for a 20 day month and model $j = 1$ receives 6.67 percent of the orders. Models $j = 2$, 3, 4 and 5 has 13.3, 20.0, 26.7 and 33.3 %, respectively. On average, the mold stays in the shoe for 1.25 days in the manufacturing process. The plant management wants to know how many molds to have in the plant inventory by model to achieve service levels between 85 and 99 %. See the Table 10.10.

Note, for model $j = 1$, the 20 days forecast is F20 = 0.067 × 1200 = 80 pair. The associated one day forecast becomes F1 = 1/20F20 = 4 pair. The average process time for all the models is $\tau_s = 1.25$ days, and therefore, $\mu = 1/1.25 = 0.80$ per day. Hence, $\lambda = 4$ per day, and because $\mu = 0.8$, $\rho = 5.0$. Using the results from Table 10.9, the minimum number of molds needed (k) by SL are 9, 9, 10, and 12. In a corresponding way, the minimum number of lasts needed for models $j = 1$, 2, 3, 4, and 5 are shown in Table 10.10. The sum of the five lasts needed range from 103 to 128. If the SL is set at 0.90, the results show where the number of lasts to have in the plant inventory is 9, 16, 22, 27 and 33 for lasts $j = 1$, 2, 3, 4 and 5, respectively.

Table 10.9 Minimum k to achieve the SL for utilization ratio ρ

	SL			
	0.85	0.90	0.95	0.99
ρ	k			
0.1	1	1	2	2
0.2	2	2	2	3
0.3	2	2	2	3
0.4	2	2	3	3
0.5	2	2	3	4
0.6	2	3	3	4
0.7	3	3	3	4
0.8	3	3	4	4
0.9	3	3	4	5
1	3	3	4	5
2	5	5	6	7
3	6	6	7	9
4	7	8	9	10
5	9	9	10	12
10	15	16	17	19
15	21	22	23	26
20	26	27	29	32
25	32	33	35	39
30	38	39	41	45
35	43	44	47	51
40	49	50	52	57
45	54	56	58	63
50	60	61	64	66
55	65	67	69	74
60	70	72	75	80
65	76	78	80	86
70	81	83	86	92
75	86	88	91	97
70	92	94	97	103
85	97	99	102	109
90	102	105	108	114
95	108	110	113	120
100	113	115	119	125
200	218	221	226	235
300	322	326	331	343
400	425	429	436	449
500	528	533	540	555
600	631	636	644	660
700	733	739	747	765

Table 10.10 Model j, probability of use p, 20 day forecast F20, 1 day forecast F1, arrival rate λ, service rate μ, utilization ratio ρ, and minimum k to achieve SL

							SL			
j	p	F20	F1	λ	μ	ρ	0.85	0.90	0.95	0.99
1	0.067	80	4	4	0.8	5.0	9	9	10	12
2	0.133	160	8	8	0.8	10.0	15	16	17	19
3	0.200	240	12	12	0.8	15.0	21	22	23	26
4	0.267	320	16	16	0.8	20.0	26	27	29	32
5	0.333	400	20	20	0.8	25.0	32	33	35	39
Sum	1.000	1200	60				103	107	114	128

Summary

Some units of product are best assembled by a lone operator, like in a pair of shoes where the exact style shape, size, and width are needed in the production process. The bill-of-material identifies the parts needed for each pair scheduled. Some of the parts are unique by style and size, and others are common between two of more varieties. Prior to the assembly date, the management determines the style and size needs for the planning horizon. The exact inventory of parts is needed to enable the workers to complete the assembly tasks. This type of assembly sometimes requires a mold to carry out the needs for each unit in production. This calls for the plant to have an inventory of the molds in storage to be used as needed by the workers. The chapter shows how to determine the size of the mold inventory to allow the plant to run in an efficient manner.

Chapter 11
Similarity Index

Introduction

This chapter pertains to a mixed model assembly plant, and measures the similarity between the models based on the work elements. Two types of measures are developed, the utilization index and the similarity index. Both indices are measures of the similarity between the models. The latter index ranges from 0 to 1, where zero occurs when there is no similarity and one is when there is full similarity. The indices are developed from the work elements where some of the elements are common to all of the models, some are unique to a particular model and others are common to two or more models. Three scenarios are described: (1) where the elements and model usage are (0 or 1) and 1 indicates the element does apply with the model; (2) where the elements are (0 or t_e) and t_e is the common element time to all models where the element applies; and (3) where the elements are (0 and t_{ej}) and t_{ej} is the time for element *e* and model *j*. The indices can be measured for sets of two or more model combinations. Examples are given to illustrate how the similarity index may be used in assembly planning.

Three Scenarios

The chapter concerns the similarity of the models that are to be produced in a plant. The similarity is measured from the work elements and how they are associated with the models for each of the three scenarios. For convenience, the element measures for the three scenarios are listed as *x*(*e, j*) for element *e* and model *j*. The scenarios are described below:

(1) Treat all $t_{ej} > 0$ with equal weight.

$$\begin{aligned} x(e,j) &= 0 \text{ if } t_{ej} = 0 \\ &= 1 \text{ if } t_{ej} > 0 \end{aligned}$$

N. T. Thomopoulos, *Assembly Line Planning and Control*,
DOI: 10.1007/978-3-319-01399-2_11,

(2) This is when all $t_{ej} > 0$ are equal, and therefore, $t_e = t_{ej}$.

$$\begin{aligned} x(e,j) &= 0 \quad \text{if } t_{ej} = 0 \\ &= t_e \quad \text{if } t_{ej} > 0 \end{aligned}$$

(3) This is when all $t_{ej} > 0$ are not necessarily equal.

$$\begin{aligned} x(e,j) &= 0 \quad \text{if } t_{ej} = 0 \\ &= t_{ej} \quad \text{if } t_{ej} > 0 \end{aligned}$$

Model Sets

A set, denoted as j^*, is defined here as a group of two or more models. For example, if models $j = 1$ and 2 are grouped together, the set is $j^* = (1,2)$. If $j = 1$, 2, and 3 are grouped, the set is $j^* = (1,2,3)$, and so forth.

The number of models in the set j^* is labeled as $N(j^*)$. Note, $N(j^*) = 2$ for $j^* = (1,2)$, and $N(j^*) = 3$ for $j^* = (1,2,3)$.

For the set j^*, the following notation is also used,

$$x'(e) = \max[x(e,j)\ (j \in j^*)]$$

Utilization Index

For the set j^*, a measure called the utilization index, $U(j^*)$ is now measured for all of the combination of sets and over all of the Ne elements. This measure is the following;

$$U(j^*) = \sum_e \sum_j x(e,j) \Big/ \left[N(j^*) \sum_e x'(e)\right]$$

The range of the utilization index will fall somewhere between $1/N(j^*)$ and 1.0, as shown below:

$$1/N(j^*) \le U(j^*) \le 1$$

When all of the elements are the same for all the models in set j^*, $U(j^*) = 1.0$. When none are the same, $U(j^*) = 1/N(j^*)$.

Similarity Index

Another measure of the relation of the elements in set j^* is called the similarity index, denoted as $S(j^*)$. This is measured as below:

$$S(j^*) = [U(j^*) - 1/N(j^*)]/[1 - 1/N(j^*)]$$

Note, where the similarity index falls between zero and one, as shown below,

$$0 \leq S(j^*) \leq 1$$

When all of the elements are the same for all the models in set j^*, $S(j^*) = 1$. When none are the same, $S(j^*) = 0$. The closer $S(j^*)$ is to one, the more similar the elements are among the models.

Scenario 1

Table 11.1 is a list of the model usage data for 10 elements and three models. The entries are set to one when element e is used on model j, and zero otherwise.

The utilization and similarity indices—rounded to two decimal places—are listed in Table 11.2.

Note when the set is $j^* = (1, 2)$, $N(j^*) = 2$ represents the number of models in the set. The maximum value for each element is computed by: $x'(e) = \max(x(e,1),\ x(e,2))$. The utilization index becomes,

$$\begin{aligned} U(j^*) &= \left[\sum\nolimits_e \sum\nolimits_j x(e,j)\right] / \left[N(j^*) \times \sum\nolimits_e x'(e)\right] \\ &= [11]/[2 \times 8] = 0.6875 \end{aligned}$$

Table 11.1 Model usage, u_{ej}, of element, e, and model, j

	j		
e	1	2	3
1	0	1	1
2	1	1	1
3	1	0	0
4	0	1	1
5	0	0	1
6	0	0	1
7	0	1	1
8	0	1	1
9	1	1	1
10	1	1	1

Table 11.2 Scenario 1, utilization, U, and similarity, S, indices for sets j^*

j^*			U	S
1	2		0.69	0.37
1	3		0.65	0.30
2	3		0.89	0.78
1	2	3	0.67	0.50

The associated similarity index is

$$S(j^*) = [U(j^*) - 1/N(j^*)]/[1 - 1/N(j^*)]$$
$$= \left[0.6875 - {}^1\!/_2\right] / \left[1 - {}^1\!/_2\right] = 0.375$$

When $j^* = (1, 2, 3)$, $N(j^*) = 3$ and $x'(e) = \max[x(e, 1), x(e, 2), x(e, 3)]$. The utilization index is

$$U(j^*) = [20]/[3 \times 10] = 0.667$$

The corresponding similarity index is

$$S(j^*) = [0.667 - 1/3]/[1 - 1/3] = 0.50$$

Scenario 2

The element data by model is listed in Table 11.3. The data contain the element times for each of the 10 elements and the 3 models. The element time is the same for all models where the same element is used.

The utilization and similarity indices—rounded to two decimal places—are listed in Table 11.4. Below shows how the utilization and similarity indices are computed.

Table 11.3 Element time, t_e, for element, e, and model, j

	j		
e	1	2	3
1	3	3	3
2	0	1	1
3	0	3	3
4	0	3	0
5	3	0	3
6	0	2	2
7	2	2	2
8	1	0	1
9	0	0	2
10	0	3	3

Table 11.4 Scenario 2, utilization, U, and similarity, S, indices for sets j^* when weighted by element times, t_e

j^*			U	S
1	2		0.62	0.24
1	3		0.72	0.45
2	3		0.80	0.61
1	2	3	0.67	0.50

When the set is $j^* = (1, 2)$ the utilization index is the following:

$$\begin{aligned} U(j^*) &= \left[\sum_e \sum_j x(e,j)\right] \Big/ \left[N(j^*) \times \sum_e x'(e)\right] \\ &= [26]/[2 \times 21] = 0.619 \end{aligned}$$

The associated similarity index is

$$\begin{aligned} S(j^*) &= [\,U(j^*) - 1/N(j^*)\,]/[1 - 1/N(j^*)] \\ &= [0.619 - {}^1\!/\!_2]/[1 - {}^1\!/\!_2] = 0.238 \end{aligned}$$

When $j^* = (1, 2, 3)$,

$$U(j^*) = [46]/[3 \times 23] = 0.667$$

The corresponding similarity index is

$$S(j^*) = [0.667 - 1/3]/[1 - 1/3] = 0.50$$

Scenario 3

The element data by model is listed in Table 11.5. The data contain the element times for each of the 10 elements and the 3 models. In this situation, the element times are not always the same for all models where the same element is used. Note,

Table 11.5 Element time, t_e, for element, e, and model, j

	j		
e	1	2	3
1	3	3	3
2	0	1	1
3	0	3	3
4	0	3	0
5	3	0	3
6	0	2	2
7	2	2	2
8	1	0	1
9	0	0	2
10	0	3	3

Table 11.6 Scenario 3, utilization, U, and similarity, S, indices for sets j^* when weighted by element times, t_{ej}

j^*			U	S
1	2		0.62	0.24
1	3		0.67	0.35
2	3		0.76	0.52
1	2	3	0.64	0.46

for example, element $e = 1$, has element times of 3, 3, and 2, for models = 1, 2, and 3, respectively.

The utilization and similarity indices—rounded to two decimal places—are listed in Table 11.6. Below shows how the utilization and similarity indices are computed.

When the set is $j^* = (2,3)$, the utilization index is the following:

$$U(j*) = \left[\sum\nolimits_e \sum\nolimits_j x(e,j)\right] \Big/ \left[N(j^*) \times \sum\nolimits_e x'(e)\right]$$
$$= [35]/[2 \times 23] = 0.761$$

The associated similarity index is

$$S(j^*) = [U(j^*) - 1/N(j^*)]/[1 - 1/N(j^*)]$$
$$= [0.761 - 1/2]/[1 - 1/2] = 0.522$$

When $j^* = (1,2,3)$,

$$U(j^*) = [44]/[3 \times 23] = 0.638$$

The corresponding similarity index is

$$S(j^*) = [0.638 - 1/3]/[1 - 1/3] = 0.458$$

Example of 100 Elements and Six Models

Table 11.7 is a list of the first 25 elements from a 100 element example. For brevity, not all the elements are shown. The data is the index of (0,1) where, the index is zero when element e is not used on model j, and is set to one, when it is used on model j.

Table 11.8 contains all combination of utilization and similarity indices for the 100 element example. The model sets are for $N(j^*) = 2$ models all the way to $N(j^*) = 6$ models.

Example 11.1 The two model similarity indices can be applied in batch assembly plants to determine the best sequence of models to send down the line in batch sizes. The best arrangement is to have the higher similarity indices apply. For example, consider the choice of the sequence as follows:

Table 11.7 A list of the first 25 of 100 elements of the model usage for element *e* and model *j*

	j					
e	1	2	3	4	5	6
1					1	1
2			1	1	1	1
3				1	1	1
4	1	1	1	1	1	1
5	1	1	1	1	1	1
6						1
7	1	1	1	1	1	1
8						1
9						1
10					1	1
11	1	1	1	1	1	1
12		1	1	1	1	1
13						1
14						1
15	1	1	1	1	1	1
16						1
17						1
18	1	1	1	1	1	1
19						1
20	1	1	1	1	1	1
21			1	1	1	1
22						1
23	1	1	1	1	1	1
24				1	1	1
25			1	1	1	1

Batch sequence: 1, 2, 3, 4, 5, 6, 1 has $S = 0.80, 0.87, 0.85, 0.88, 0.75, 0.56$, respectively.

Another arrangement of the batch sequence is listed along with the corresponding two model similarity indices:

Batch sequence: 3, 1, 5, 4, 2, 6, 3 has $S = 0.70, 0.52, 0.88, 0.74, 0.49, 0.56$, respectively.

Since the average S is 0.785 for the first sequence, and 0.648 for the second sequence, the first sequence is preferred.

Example 11.2 Suppose the plant with six models will run two lines with three models on each line, and the three options given are considered.

$j^* = 1,2,3$ with $S = 0.79$ on line 1 and $j^* = 4,5,6$ with $S = 0.70$ on line 2.
$j^* = 1,4,6$ with $S = 0.52$ on line 1 and $j^* = 2,3,5$ with $S = 0.70$ on line 2.
$j^* = 1,2,6$ with $S = 0.44$ on line 1 and $j^* = 3,4,5$ with $S = 0.81$ on line 2.

The best arrangement appears as $j^* = (1,2,3)$ in line 1 and $j^* = (4,5,6)$ in line 2, since the combined average of similarity indices (0.74, 0.61, and 0.62) is the largest among the three candidates.

Table 11.8 Utilization, U, and similarity, S, indices for model sets, j^*

j^*				U	S
1	2			0.90	0.80
1	3			0.85	0.70
1	4			0.80	0.59
1	5			0.76	0.52
1	6			0.69	0.56
2	3			0.94	0.87
2	4			0.87	0.74
2	5			0.83	0.65
2	6			0.74	0.49
3	4			0.92	0.85
3	5			0.87	0.75
3	6			0.78	0.56
4	5			0.94	0.88
4	6			0.83	0.66
5	6			0.87	0.75
1	2	3		0.86	0.79
1	2	4		0.78	0.67
1	2	5		0.72	0.59
1	2	6		0.63	0.44
1	3	4		0.81	0.72
1	3	5		0.76	0.63
1	3	6		0.65	0.47
1	4	5		0.80	0.70
1	4	6		0.68	0.52
1	5	6		0.71	0.57
2	3	4		0.86	0.80
2	3	5		0.80	0.70
2	3	6		0.68	0.52
2	4	5		0.84	0.77
2	4	6		0.72	0.57
2	5	6		0.75	0.62
3	4	5		0.88	0.81
3	4	6		0.74	0.61
3	5	6		0.77	0.65
4	5	6		0.80	0.70
1	2	3	4	0.80	0.73
1	2	3	5	0.73	0.64
1	2	3	6	0.61	0.48
1	2	4	5	0.76	0.68
1	2	4	6	0.63	0.51

(continued)

Table 11.8 (continued)

*j**						*U*	*S*
1	2	5	6			0.66	0.54
1	3	4	5			0.79	0.72
1	3	4	6			0.65	0.54
1	3	5	6			0.67	0.57
1	4	5	6			0.70	0.60
2	3	4	5			0.82	0.76
2	3	4	6			0.68	0.57
2	3	5	6			0.70	0.60
2	4	5	6			0.72	0.63
3	4	5	6			0.74	0.66
1	2	3	4	5		0.76	0.70
1	2	3	4	6		0.62	0.52
1	2	3	5	6		0.64	0.55
1	2	4	5	6		0.66	0.57
1	3	4	5	6		0.67	0.59
2	3	4	5	6		0.69	0.61
1	2	3	4	5	6	0.64	0.57

Summary

This chapter concerns a mixed model make-to-stock assembly line where a variety of models are produced. The work elements of each model and the element commonality between the models are used to measure the similarity among the models. Two indices are measured, the utilization index and the similarity index. The indices are measured for sets of two or more models at a time. Some applications of the indices are suggested. These include batch assembly where the similarity indices determine how to sequence the models, in batches, down the line. Another application concerns mixed model make-to-stock assembly where two or more lines are available. The similarity indices are used to select the combination of models to assign on each of the lines.

Chapter 12
Learning Curves

Introduction

Learning Curves can be used to estimate the time required to complete a selected number of units on an assembly line. The chapter shows how to apply learning curves for single model lines and for mixed model lines, and describes the learning rate, the learning coefficient, and the learning multiplier and how they are used to develop the learning curve. The learning curve and the unit standard time are combined to compute the learning limit for the product. The unit time is higher than the standard time for all assemblies prior to the learning limit, and for those after the learning limit, the assembly unit time is the same as the standard time. With all this information, the projected average time (and the cumulative total time) to complete a selected number of units can be computed. The chapter shows how to extend the method of learning curves for mixed model lines. Examples are given for a single model assembly line, for a 2-model assembly line, and for a 3-model assembly line. The method extends to an M-model assembly line.

When an operator on an assembly line begins a new task, the first units worked on take longer than subsequent units. The amount of time required to complete a given task will be less each time the task is undertaken and the time per unit will decrease at a decreasing rate. The reduction in time will follow a pattern that is called a learning curve.

The formulation of learning curves is based upon the power function offered by T. P. Wright in 1936. The theory states that the assembly time per unit declines by some constant percentage every time the number of assemblies is doubled. This is represented mathematically by a two parameter function. If r is the rth repetition and $t(r)$ is the time required to assemble the rth unit, then

$$t(r) = ar^b \quad r = 1, 2, \ldots$$

where

a = assembly time for the first unit,
b = a negative constant,

N. T. Thomopoulos, *Assembly Line Planning and Control*,
DOI: 10.1007/978-3-319-01399-2_12,

r^b = portion of 'a' (time of first unit) for the rth repetition.

100R represents the percentage decrease in $t(r)$ every time r is doubled. Therefore

$$t(2r)/t(r) = R \quad r = 1, 2, \ldots$$

for $0.5 < R < 1$ and R is called the learning rate. Should $R = 0.90$, say, $t(2r) = 0.9t(r)$.

Note the relation between R and b,

$$R = a(2r)^b/ar^b = 2^b$$

Below shows how to find the value of b that corresponds to R,

$$b = \ln(R)/\ln(2)$$

where ln is the natural logarithm.

For example, should the assembly time per unit decrease by 90 % each time the number of assemblies is doubled, $R = 0.90$ and $b = \ln(0.90)/\ln(2) = -0.152$.

The time required to assemble the first r units is obtained by

$$T(r) = \sum_{x=1}^{r} t(x) = \sum_{x=1}^{r} ax^b$$

When r is a large number, say larger than 100, $T(r)$ may be approximated by

$$T(r) \cong \int_0^r t(x)dx = ar^{(b+1)}/(b+1)$$

The average time for the first r units, denoted as $A(r)$, is obtained as below:

$$A(r) = T(r)/r$$

Note also where $b > -1.00$ and the learning rate is limited to the range: $0.50 < R < 1.00$.

Single Model Assembly

For assembly lines, the parameter 'a' is generally set as a multiple of the unit standard time, Σt_e. For notational convenience here, $Te = \Sigma t_e$. Hence, $a = kTe$ where k is a multiplier constant and $k \geq 1$.

In the example where the learning rate is $R = 90$ % and $a = 1.00$, the theoretical assembly times, rounded to two decimals, [$t(r)$, $T(r)$, $A(r)$] for the first eight units are listed in table.

r	t(r)	T(r)	A(r)
1	1.00	1.00	1.00
2	0.90	1.90	0.95
3	0.85	2.75	0.92
4	0.81	3.56	0.89
5	0.78	4.34	0.87
6	0.76	5.10	0.85
7	0.74	5.84	0.83
8	0.73	6.57	0.82

The following list shows the assembly times, $t(r)$, for the first eight repetitions when the learning rate is $R = 0.90$, and when the number of repetitions double, i.e., $r = (1, 2, 4, 8)$. Note how $t(2r) = R \times t(r)$.

r	t(r)
1	1.00
2	0.90
4	0.81
8	0.73

Learning Limit

In theory, the assembly time continues to decrease by a rate of R every time the number of repetitions double. But, in actuality, the assembly time reaches a limit where the time ceases to decrease. One way to set the limit is to find the repetition where the projected assembly time is the same as the unit time, Σt_e. That would be when the assembly time is $t(r) = Te$. The number of repetitions when this limit occurs is here called the learning limit and is denoted by r_o. To find the learning limit, r_o, note the relation below:

$$t(r) = ar^b = k(Te)r^b$$

where the learning limit r_o is obtained when $t(r_o) = Te$, thereby.

$$kr_o^b = 1$$

and solving for r_o, yields,

$$r_o = (1/k)^{1/b}$$

In this way, learning continues for each repetition until $r = r_o$. Afterward, the assembly time per unit is the same as the sum of the element times, Σt_e. Hence, the assembly time, $t(r)$, for the rth repetition becomes,

$$\begin{aligned} t(r) &= ar^b \quad && r = 1 \text{ to } r_o \\ &Te && r > r_o \end{aligned}$$

In Table 12.1, selective values of R (0.95–0.60) and k (1.5, 2.0, 2.5, 3.0) are used to show the values of b, r_o, $T(r_o)$ and $A(r_o)$ in a generic unit of time. For simplicity, the table assumes the standard unit time, Te, is 1.00. Note, $T(r_o)$ is the total time needed to complete r_o units when the learning rate is R and the multiplier is k. Also, $A(r_o)$ is the average time needed for the first r_o units.

Note when $R = 0.80$, $k = 2.0$ and $b = -0.322$,

$$\begin{aligned} r_o &= (1/k)^{(1/b)} \\ &= (1/2)^{(1/-0.322)} \\ &= 8.610. \end{aligned}$$

The table value rounds $r_o = 9$.

Note when $R = 0.90$ and $k = 2.00$, $r_o = 96$. The time to complete the first 96 units is $T(96) = 112$, and the average time per unit is $A(96) = 1.171$.

In general, the projected assembly time, $t(r)$, for any value of r, is as follows:

$$\begin{aligned} t(r) &= kTe\, r^b \quad && \text{for } r \leq r_0 \\ &Te && \text{for } r > r_0 \end{aligned}$$

The corresponding total time, $T(r)$, is shown below:

$$T(r) = \sum_{x=1}^{r} t(x)$$

Further, the average time per unit is obtained as below:

$$A(r) = T(r)/r$$

Note, $A(r)$ represents the ratio of the average (learning) time over the standard time, Te.

Table 12.2 contains the average time, $A(r)$, for the selected values of R and k and for repetitions going from 10 to 100. On a second page, the repetitions continue from 100 to 1,000. Recall, the table is based on $Te = 1.00$. The average times are computed as described earlier. Note, when $R = 0.80$, $k = 2.0$ and $r = 50$,

$$A(50) = 1.054.$$

To compute the average time when $r > r_o$, some entries from Table 12.1 are needed. Note at $R = 0.80$ and $k = 2.0$, the table lists the rounded value of $r_o = 9$ and also $A(ro) = 1.313$. The non-rounded computation is $r_o = 8.610$. The following shows how to find the average time for $r = 100$, say:

Table 12.1 Learning rate, R, coefficient, b, multiplier, k, learning limit, r_o, total time, $T(r_o)$, and average time, $A(r_o)$

R	b	k	r_o	$T(r_o)$	$A(r_o)$
0.95	−0.074	1.5	240	258	1.080
0.95	−0.074	2.0	11,693	12,627	1.080
0.95	−0.074	2.5	238,520	257,581	1.080
0.95	−0.074	3.0	2,80,2424	3,026,377	1.080
0.90	−0.152	1.5	14	16	1.145
0.90	−0.152	2.0	96	112	1.171
0.90	−0.152	2.5	415	488	1.178
0.90	−0.152	3.0	1,377	1,622	1.178
0.85	−0.234	1.5	6	7	1.186
0.85	−0.234	2.0	19	24	1.250
0.85	−0.234	2.5	50	64	1.277
0.85	−0.234	3.0	108	140	1.290
0.80	−0.322	1.5	4	4	1.214
0.80	−0.322	2.0	9	11	1.313
0.80	−0.322	2.5	17	24	1.366
0.80	−0.322	3.0	30	42	1.397
0.75	−0.415	1.5	3	3	1.235
0.75	−0.415	2.0	5	7	1.361
0.75	−0.415	2.5	9	13	1.441
0.75	−0.415	3.0	14	21	1.495
0.70	−0.515	1.5	2	3	1.250
0.70	−0.515	2.0	4	5	1.399
0.70	−0.515	2.5	6	9	1.503
0.70	−0.515	3.0	8	13	1.581
0.65	−0.621	1.5	2	2	1.260
0.65	−0.621	2.0	3	4	1.430
0.65	−0.621	2.5	4	7	1.560
0.65	−0.621	3.0	6	10	1.655
0.60	−0.737	1.5	2	2	1.288
0.60	−0.737	2.0	3	4	1.468
0.60	−0.737	2.5	3	6	1.609
0.60	−0.737	3.0	4	8	1.724

$$A(100) = \{8.610 \times 1.313 + 91.39 \times 1.00\}/100 = 1.027$$

Example 12.1 Suppose the ABC Corporation receives an order for 1,000 units to be produced on their assembly line with a shift time of $T = 450$ min. The engineers set the standard time for a unit as $Te = 200$ min. The learning rate is projected as $R = 0.90$ and the multiplier is $k = 2.5$. The schedule calls for about $N = 100$ units a day. The number of operators needed per day becomes,

$$n = N \times Te/T = 100 \times 200/450 = 44.44$$

Table 12.2 The average time, $A(r)$, at learning rate, R, multiplier, k, and repetitions, r

		r																
R	k	10	20	30	40	50	60	70	80	90	100	200	300	400	500	600	700	800	900	1,000
0.95	1.5	1.343	1.285	1.250	1.226	1.207	1.191	1.178	1.167	1.157	1.149	1.093	1.063	1.047	1.038	1.031	1.027	1.023	1.021	1.019
0.95	2.0	1.791	1.713	1.667	1.634	1.609	1.588	1.571	1.556	1.543	1.532	1.457	1.415	1.385	1.363	1.344	1.329	1.316	1.305	1.295
0.95	2.5	2.239	2.141	2.084	2.043	2.011	1.986	1.964	1.945	1.929	1.915	1.821	1.768	1.731	1.703	1.681	1.662	1.645	1.631	1.619
0.95	3.0	2.686	2.570	2.500	2.451	2.413	2.383	2.357	2.335	2.315	2.298	2.185	2.122	2.078	2.044	2.017	1.994	1.975	1.957	1.942
0.90	1.5	1.199	1.104	1.069	1.052	1.042	1.035	1.030	1.026	1.023	1.021	1.010	1.007	1.005	1.004	1.003	1.003	1.003	1.002	1.002
0.90	2.0	1.599	1.461	1.382	1.327	1.286	1.252	1.225	1.201	1.181	1.163	1.082	1.054	1.041	1.033	1.027	1.023	1.020	1.018	1.016
0.90	2.5	1.999	1.826	1.727	1.659	1.607	1.566	1.531	1.502	1.476	1.454	1.312	1.235	1.183	1.146	1.122	1.105	1.092	1.081	1.074
0.90	3.0	2.398	2.191	2.073	1.991	1.929	1.879	1.837	1.802	1.771	1.744	1.574	1.482	1.420	1.373	1.336	1.305	1.279	1.256	1.237
0.85	1.5	1.105	1.052	1.035	1.026	1.021	1.017	1.015	1.013	1.012	1.010	1.005	1.003	1.003	1.002	1.002	1.001	1.001	1.001	1.001
0.85	2.0	1.423	1.241	1.160	1.120	1.096	1.080	1.069	1.060	1.053	1.048	1.024	1.016	1.012	1.010	1.008	1.007	1.006	1.005	1.005
0.85	2.5	1.779	1.550	1.424	1.339	1.276	1.230	1.197	1.172	1.153	1.138	1.069	1.046	1.034	1.028	1.023	1.020	1.017	1.015	1.014
0.85	3.0	2.135	1.860	1.709	1.607	1.531	1.471	1.421	1.380	1.344	1.313	1.157	1.104	1.078	1.063	1.052	1.045	1.039	1.035	1.031
0.80	1.5	1.075	1.038	1.025	1.019	1.015	1.013	1.011	1.009	1.008	1.008	1.004	1.003	1.002	1.002	1.001	1.001	1.001	1.001	1.001
0.80	2.0	1.269	1.135	1.090	1.067	1.054	1.045	1.038	1.034	1.030	1.027	1.013	1.009	1.007	1.005	1.004	1.004	1.003	1.003	1.003
0.80	2.5	1.579	1.315	1.210	1.158	1.126	1.105	1.090	1.079	1.070	1.063	1.032	1.021	1.016	1.013	1.011	1.009	1.008	1.007	1.006
0.80	3.0	1.895	1.573	1.402	1.301	1.241	1.201	1.172	1.151	1.134	1.121	1.060	1.040	1.030	1.024	1.020	1.017	1.015	1.013	1.012
0.75	1.5	1.063	1.031	1.021	1.016	1.013	1.010	1.009	1.008	1.007	1.006	1.003	1.002	1.002	1.001	1.001	1.001	1.001	1.001	1.001
0.75	2.0	1.192	1.096	1.064	1.048	1.038	1.032	1.027	1.024	1.021	1.019	1.010	1.006	1.005	1.004	1.003	1.003	1.002	1.002	1.002
0.75	2.5	1.401	1.200	1.134	1.100	1.080	1.067	1.057	1.050	1.045	1.040	1.020	1.013	1.010	1.008	1.007	1.006	1.005	1.004	1.004
0.75	3.0	1.677	1.349	1.233	1.175	1.140	1.116	1.100	1.087	1.078	1.070	1.035	1.023	1.017	1.014	1.012	1.010	1.009	1.008	1.007
0.70	1.5	1.055	1.028	1.018	1.014	1.011	1.009	1.008	1.007	1.006	1.006	1.003	1.002	1.001	1.001	1.001	1.001	1.001	1.001	1.001
0.70	2.0	1.154	1.077	1.051	1.038	1.031	1.026	1.022	1.019	1.017	1.015	1.008	1.005	1.004	1.003	1.003	1.002	1.002	1.002	1.002
0.70	2.5	1.299	1.149	1.100	1.075	1.060	1.050	1.043	1.037	1.033	1.030	1.015	1.010	1.007	1.006	1.005	1.004	1.004	1.003	1.003
0.70	3.0	1.491	1.245	1.164	1.123	1.098	1.082	1.070	1.061	1.055	1.049	1.025	1.016	1.012	1.010	1.008	1.007	1.006	1.005	1.005

(continued)

Table 12.2 (continued)

		r																		
R	k	10	20	30	40	50	60	70	80	90	100	200	300	400	500	600	700	800	900	1,000
0.65	1.5	1.050	1.025	1.017	1.013	1.010	1.008	1.007	1.006	1.006	1.005	1.003	1.002	1.001	1.001	1.001	1.001	1.001	1.001	1.001
0.65	2.0	1.131	1.066	1.044	1.033	1.026	1.022	1.019	1.016	1.015	1.013	1.007	1.004	1.003	1.003	1.002	1.002	1.002	1.001	1.001
0.65	2.5	1.244	1.122	1.081	1.061	1.049	1.041	1.035	1.031	1.027	1.024	1.012	1.008	1.006	1.005	1.004	1.003	1.003	1.003	1.002
0.65	3.0	1.384	1.192	1.128	1.096	1.077	1.064	1.055	1.048	1.043	1.038	1.019	1.013	1.010	1.008	1.006	1.005	1.005	1.004	1.004
0.60	1.5	1.050	1.025	1.017	1.013	1.010	1.008	1.007	1.006	1.006	1.005	1.003	1.002	1.001	1.001	1.001	1.001	1.001	1.001	1.001
0.60	2.0	1.120	1.060	1.040	1.030	1.024	1.020	1.017	1.015	1.013	1.012	1.006	1.004	1.003	1.002	1.002	1.002	1.002	1.001	1.001
0.60	2.5	1.211	1.106	1.070	1.053	1.042	1.035	1.030	1.026	1.023	1.021	1.011	1.007	1.005	1.004	1.004	1.003	1.003	1.002	1.002
0.60	3.0	1.322	1.161	1.107	1.080	1.064	1.054	1.046	1.040	1.036	1.032	1.016	1.011	1.008	1.006	1.005	1.005	1.004	1.004	1.003

Table 12.1 shows when $R = 0.90$ and $k = 2.5$, the learning limit is $r_o = 415$ (repetitions) assemblies, and the associated total time is $T(415) = 488$. Hence the total time to assemble $N = 1{,}000$ units becomes,

$$\begin{aligned} T(1,000) &= T(415) + 585 \\ &= 488 + 585 \\ &= 1073 \end{aligned}$$

Table 12.2 shows $T(1{,}000) = 1{,}074$; the difference is due to rounding. Using, 1,074, the time in minutes becomes,

$$\begin{aligned} T(1000)_{\text{min}} &= 1074 \times Te \\ &= 1074 \times 200 \\ &= 214,800 \text{ min} \end{aligned}$$

and in days,

$$\begin{aligned} T(1000)_{\text{days}} &= 214,800/(T \times n) \\ &= 214,800/(450 \times 44.44) \\ &= 10.74 \text{ days} \end{aligned}$$

For notation sake, $Nd = 10.74$ represents the number of days to complete the order of $N = 1{,}000$ units.

Example 12.2 Continuing with Example 12.1, the management wants to know how many days are needed to assemble the first 100 units. Using Table 12.2 again, the average time of the first 100 units is $A(100) = 1.454$. In minutes, this is

$A(100)_{\text{min}} = 1.454 \times Te = 1.454 \times 200 = 290.8$ min
The total time becomes $T(100) = 290.8 \times 100 = 29{,}080$ min.
The number of days, becomes:
$A(100)_{\text{days}} = 29{,}080/(T \times n) = 29{,}080/(450 \times 44.44) = 1.454$ days

Example 12.3 Continuing with Example 12.1, the management wants to estimate the material and labor cost per unit for the lot size of 1,000 units. The material cost is set at $C_M = \$800$ per unit and the labor cost is $C_L = \$30$ per hour at an 8 hour day. With this information, the cost per unit becomes

$$\begin{aligned} C_{1,000} &= nC_L 8Nd + C_M 1,000 \\ &= 44.44 \times 30 \times 8 \times 10.74 + 800 \times 1,000 \\ &= \$914,548 \end{aligned}$$

whereby, the cost per unit, C_U, is

$$\begin{aligned} C_U &= C_{1,000}/1,000 = 914,548/1,000 \\ &= \$914.55 \end{aligned}$$

Example 12.4 Suppose a single model line has a shift time of $T = 450$ min, and traditionally, the learning multiplier for the first unit down the line is $k = 2.5$. A new product is run using $n =$ ten operators and has a standard unit time of $Te = 80$ min. The average cycle time is $\overline{c} = Te/n = 80/10 = 8$ min. Using the standard time, with $\overline{c} = 8.00$ min and $T = 450$ min, the number of units per shift will be $T/\overline{c} = 56.25$. Suppose $N1 = 50$ units are completed during the first shift, and the management is seeking the estimate of the learning rate, R.

Since, $T \times n/Te = 56.25$ and $T \times n/Te' = 50.00$, and Te' is the average unit time for the first 50 units, then, $Te'/Te = N/N1 = 56.25/50 = 1.125 = A(50)'$ is the average unit time over the standard time for the first $N1 = 50$ units. This is an estimate of $A(50)$ at $k = 2.5$. Table 12.2 is searched seeking the learning rate, with $k = 2.5$, that yields similar results. The table shows the following:

$$R = 0.80 \text{ has } A(50) = 1.126$$

Hence, $R \cong 0.80$.

Estimating the Learning Rate

Example 12.4 shows how to estimate the learning rate, R, for an individual product with use of the data in Table 12.2. The method described could be applied to the various products that are scheduled for assembly in the plant. As more and more estimates of the learning rate, R, become available, the plant management can establish an average and confidently apply the various timing and costs applications associated with learning curves on the assembly line of the plant.

Mixed Model Assembly

In mixed model assembly, the operators are not always performing the same tasks on all models. Some elements are unique to some models, while others are common to two or more models. As a result, the rate at which operators are learning varies from element to element and from model to model. Because of this, some modifications are needed on the mixed model learning curve

Two Model Learning Curve

Consider a two model assembly line with models $j = 1, 2$. Suppose T_1 is the standard work time for one unit of $j = 1$ and T_2 is the same for $j = 2$. The number of units per shift to produce is N_1 for $j = 1$ and N_2 for $j = 2$. $N = N_1 + N_2$ is the shift schedule quantity.

Time for First Unit in Learning

Recall in the single model learning formulation, $a = k\,\Sigma t_e$ denotes the time for the first unit. Σt_e is the standard unit time for all the elements e, and k is a learning multiplier. In mixed model learning, the time for the first unit is a weighted average over the two models as shown below:

$$a = k(N_1T_1 + N_2T_2)/N$$

Unique and Common Elements

Let j^* represent a set of models where $j^* = (1, 2, 12)$ are for the sets of $j = 1$, $j = 2$ and for $j = 1$ and 2, respectively. The set of elements associated with the three sets are denoted, $E(1)$, $E(2)$, and $E(12)$, respectively. In the same way, the standard time for the three sets are denoted as $T(1)$, $T(2)$ ant $T(12)$, respectively. The number of repetitions over a shift for the three sets is as below:

$$N(1) = N_1$$
$$N(2) = N_2$$
$$N(12) = N_1 + N_2 = N$$

The shift times devoted to $E(1)$, $E(2)$, $E(12)$ are the following: $N(1)T(1)$, $N(2)T(2)$, and $N(12)T(12)$. The total time for the shift is

$$\sum N_jT_j = N_1T_1 + N_2T_2 = N(1)T(1) + N(2)T(2) + N(12)T(12)$$

So now, the portion of the shift time devoted to the three sets are as below:

$$P(1) = N(1)T(1)/\sum N_jT_j$$
$$P(2) = N(2)T(2)/\sum N_jT_j$$
$$P(12) = N(12)T(12)/\sum N_jT_j$$

Repetitions

In the learning formulation, $t(r) = ar^b$, recall, r^b represents the portion of time from the first unit, 'a', that is needed to complete the rth repetition. In mixed model, the number of repetitions varies by the three sets as shown below:

r^b = the fraction of 'a' on the rth unit.
$(rN(1)/N)^b$ = fraction of 'a' for an $E(1)$ element on the rth unit.
$(rN(2)/N)^b$ = fraction of 'a' for an $E(2)$ element on the rth unit.
$(rN(12)/N)^b$ = fraction of 'a' for an $E(12)$ element on the rth unit.

Since $N(12) = N$, $N(12)/N = 1.0$, and thereby, r^b = fraction of 'a' for an $E(12)$ element on the rth unit.

In general, the portion of time for the rth unit becomes:

$$\begin{aligned} r^b &= P(1)[r(N1)/N)]^b + P(2)\left[r(N2)/N\right]^b + P(12)r^b \\ &= r^b\left\{P(1)[N(1)/N]^b + P(2)[N(2)/N]^b + P(12)\right\} \\ &= r^b Q \end{aligned}$$

Q is a coefficient greater than one that represents the added learning multiplier due to the mixed model elements and schedules.

Mixed Model Learning Curve

The mixed model learning curve for this two model line is the following:

$$t(r) = ar^bQ \qquad r = 1, 2, \ldots$$

Example 12.5 Suppose the two model line with schedule quantities for models 1 and 2 are $N_1 = 20$ and $N_2 = 30$; altogether $N = 50$ units for the shift. The standard unit time for models 1 and 2 are $T_1 = 80$ and $T_2 = 70$ min. The learning rate is set at $R = 0.90$ and the multiplier for the first unit is $k = 2.5$. Recall when $R = 0.90$, $b = -0.152$. The mixed model adjustment for the first unit is computed as below:

$$\begin{aligned} a &= k[N_1T_1 + N_2T_2]/N \\ &= 2.5[20 \times 80 + 30 \times 70]/50 \\ &= 185 \text{ min} \end{aligned}$$

The data for the three set of elements are listed in Table 12.3. Note, the standard time for $j^* = 1$ and 2 are 20 and 10 min, respectively.

The mixed model learning coefficient, Q, is computed as below:

Table 12.3 Element sets, $E(j^*)$, standard time $T(j^*)$, shift schedule $N(j^*)$, product, $N(j^*)T(j^*)$, and Probability, $P(j^*)$

E(j*)	T(j*)	N(j*)	N(j*)/N	N(j*)T(j*)	P(j*)
$E(1)$	20	20	0.4	400	0.108
$E(2)$	10	30	0.6	300	0.081
$E(12)$	60	50	1.0	3,000	0.811
Sum				3,700	1.000

$$\begin{aligned} Q &= \left\{P(1)[N(1)/N]^b + P(2)[N(2)/N]^b + P(12)\right\} \\ &= \{0.108[0.40]^{-0.152} + 0.081[0.60]^{-0.152} + 0.811\} \\ &= 1.023 \end{aligned}$$

Hence, the mixed model learning curve becomes:

$$\begin{aligned} t(r) &= ar^bQ \\ &= 185r^{-0.152} \times 1.023 \end{aligned}$$

Three Model Learning Curve

Example 12.6 Now consider a three model line with models $j = 1, 2, 3$. The schedule over a shift is $N_1 = 50$, $N_2 = 30$, and $N_3 = 20$ yielding $N = 100$ units altogether. The standard unit times for the models are: $T_1 = 20.4$, $T_2 = 29.6$, and $T_3 = 33.9$ min. The learning rate is set at $R = 0.90$ and the learning multiplier is $k = 2.00$. The time for the first unit in learning becomes:

$$\begin{aligned} a &= 2.0[50 \times 20.4 + 30 \times 29.6 + 20 \times 33.9]/100 \\ &= 51.66 \end{aligned}$$

Table 12.4 is a list of all the elements, e, and their element times, t_e. The usage index of the element e by model j, u_{ej} is also listed for each element.

The mixed model learning coefficient, Q, (Table 12.5) becomes:

$$\begin{aligned} Q = [&0.033 \times 0.5^b + 0.061 \times 0.3^b + 0.039 \times 0.2^b + 0.084 \times 0.8^b \\ &+ 0.195 \times 0.7^b + 0.247 \times 0.5^b + 0.341] \end{aligned}$$

Since, the learning rate is $R = 0.90$, $b = -0.152$, and thereby,

$$Q = 1.068.$$

Table 12.4 Element, *e*, element time, t_e, and usage index, u_{ej}, for model *j*

		u_{ej}		
e	t_e	1	2	3
1	2.4	1	1	1
2	1.9	1	0	1
3	3.2	1	1	1
4	0.7	1	0	1
5	1.9	0	1	1
6	0.8	1	0	0
7	1.5	1	0	1
8	2.2	0	1	1
9	0.4	1	0	1
10	0.9	1	1	1
11	1.4	1	0	1
12	2	0	1	0
13	1.3	1	0	1
14	0.9	1	0	0
15	3.3	0	1	0
16	1.6	0	1	1
17	1.3	0	0	1
18	1.5	1	1	0
19	2.2	0	1	1
20	1.6	0	0	1
21	1.2	1	1	0
22	2.5	0	1	1
23	2.3	1	1	1
24	2.4	0	1	1
25	2.2	0	0	1

Table 12.5 is a list of all the element sets, $E(j^*)$, along with the standard times, $T(j^*)$, shift repetitions, $N(j^*)$, $N(j^*)/N$, $N(j^*)T(j^*)$, and $P(j^*)$

E(j*)	T(j*)	N(j*)	N(j*)/N	T(j*)N(j*)	P(j*)
E(1)	1.7	50	0.5	85	0.033
E(2)	5.3	30	0.3	159	0.061
E(3)	5.1	20	0.2	102	0.039
E(12)	2.7	80	0.8	216	0.084
E(13)	7.2	70	0.7	504	0.195
E(23)	12.8	50	0.5	640	0.247
E(123)	8.8	100	1.0	880	0.341
Sum					1.000

The mixed model learning curve becomes:

$$\begin{aligned} t(r) &= ar^b Q \\ &= 51.66r^{-0.152} \times 1.068 \end{aligned}$$

M-Model Learning Curve

An assembly line with M models is now considered where the number of units going down the line in a shift is N_j for model j and is $N = \Sigma N_j$ for all models. The standard work time for model j is T_j when k is the learning multiplier, the weighted average time for the first unit is obtained by,

$$a = k\sum\nolimits_j N_jT_j/N = k\overline{T}j$$

where $\overline{T}j$ is the weighted average unit time for all models.

The mixed model learning coefficient, Q, is computed as follows:

$$Q = \sum\nolimits_j P(j*)[N(j*)/N]^b$$

where

$$P(j*) = [T(j*)N(j*)]/\sum\nolimits_{j*}[T(j*)N(j*)]$$

The mixed model learning curve is below:

$$t(r) = ar^bQ \quad r = 1, 2, \ldots$$

Example 12.7 An order comes in for 1,000 units covering three models to be assembled on a mixed model line. The learning rate is estimated as $R = 0.90$, the learning multiplier is $k = 2.5$, and the mixed model learning coefficient is computed as $Q = 1.05$. The shift time is $T = 450$ min, and the weighted average standard unit time over the three models is $\overline{T}j = 160$ min. Assume the number of operators on the line is $n = 15$. The management wants to estimate the number of days needed to complete the order of $N = 1{,}000$ units.

Using Table 12.6, with $R = 0.90$, $k = 2.5$, and $Q = 1.05$, the learning limit is $r_o = 572$ and $Ar_o = A(r_o) = 1.178$. Hence, the total time becomes,

$$\begin{aligned} N_{\min} &= [r_o \times A(r_o) + (1000 - r_o) \times 1] \times \overline{T}j \\ &= [572 \times 1.178 + (1000 - 572) \times 1] \times 160 \\ &= 176{,}290 \text{ min} \end{aligned}$$

The number of days becomes,

$$\left\{ \begin{aligned} N_{\text{days}} &= 176{,}290/[n \times T] \\ &= 176{,}290/[15 \times 450] \\ &= 26.12 \text{ days.} \end{aligned} \right\}$$

Table 12.6 Learning limit, r_o, at learning rate, R, multiplier, k, single model average time, Ar_o, mixed model learning coefficient, Q. * indicates $r_o > 100{,}000$

			Q								
R	k	Ar0	1.000	1.025	1.050	1.075	1.100	1.1255	1.150	1.175	1.200
0.95	1.5	1.080	240	335	463	637	869	1,177	1,584	2,119	2,816
0.95	2.0	1.080	1,1693	16,325	22,609	31,073	42,393	57,436	77,299	*	*
0.95	2.5	1.080	*	*	*	*	*	*	*	*	*
0.95	3.0	1.080	*	*	*	*	*	*	*	*	*
0.90	1.5	1.145	14	17	20	23	27	31	36	42	48
0.90	2.0	1.171	96	112	132	154	179	207	240	276	317
0.90	2.5	1.178	415	488	572	668	777	901	1,041	1,199	1,377
0.90	3.0	1.178	1,377	1,620	1,898	2,216	2,578	2,988	3,453	3,978	4,569
0.85	1.5	1.186	6	6	7	8	8	9	10	11	12
0.85	2.0	1.250	19	21	24	26	29	32	35	38	42
0.85	2.5	1.277	50	55	61	68	75	82	90	99	108
0.85	3.0	1.290	108	120	133	148	163	179	197	216	236
0.80	1.5	1.214	4	4	4	4	5	5	5	6	6
0.80	2.0	1.313	9	9	10	11	12	12	13	14	15
0.80	2.5	1.366	17	19	20	22	23	25	27	28	30
0.80	3.0	1.397	30	33	35	38	41	44	47	50	53
0.75	1.5	1.235	3	3	3	3	3	4	4	4	4
0.75	2.0	1.361	5	6	6	6	7	7	7	8	8
0.75	2.5	1.441	9	10	10	11	11	12	13	13	14
0.75	3.0	1.495	14	15	16	17	18	19	20	21	22
0.70	1.5	1.250	2	2	2	3	3	3	3	3	3
0.70	2.0	1.399	4	4	4	4	5	5	5	5	5
0.70	2.5	1.503	6	6	7	7	7	7	8	8	8
0.70	3.0	1.581	8	9	9	10	10	11	11	12	12
0.65	1.5	1.260	2	2	2	2	2	2	2	2	3
0.65	2.0	1.430	3	3	3	3	4	4	4	4	4
0.65	2.5	1.560	4	5	5	5	5	5	5	6	6
0.65	3.0	1.655	6	6	6	7	7	7	7	8	8
0.60	1.5	1.288	2	2	2	2	2	2	2	2	2
0.60	2.0	1.468	3	3	3	3	3	3	3	3	3
0.60	2.5	1.609	3	4	4	4	4	4	4	4	4
0.60	3.0	1.724	4	5	5	5	5	5	5	6	6

Example 12.8 Using the same data as Example 12.6, the management wants to estimate the number of days to complete the first 100 units. To accomplish, the computations are below.

Using $R = 0.90$, $k = 2.5$, and $r = 100$, Table 12.2 shows $A_{100} = 1.454$ for a single model line. The corresponding time for a mixed model line with $Q = 1.05$ becomes,

$$A'_{100} = Q \times A_{100}$$
$$= 1.05 \times 1.454$$
$$= 1.527$$

So, the number of minutes required, becomes,

$$\mathrm{N_{min}} = \mathrm{A'_{100}} \times \overline{T}j \times 100$$
$$= 1.527 \times 160 \times 100$$
$$= 24,432$$

The number of days is computed as below:

$$N_{\text{days}} = 24,432/(n \times T)$$
$$= 24,432/(15 \times 450)$$
$$= 3.62$$

Summary

Learning curves are used to estimate the time and cost to complete a lot size job on an assembly line. The data needed are the learning rate, the standard time per unit, and a multiplier for the first unit. The chapter shows how to determine the learning limit that identifies the number of repetitions (assemblies) until the learning process ends. The projection of the time for any repetition of assembly is computed, as well as the cumulative total time and the corresponding average time. The learning curve methods extend for mixed model make-to-stock lines. Examples show how to apply the methods for single model lines and for mixed model make-to-stock lines.

Bibliography

Sorensen CE, Williamson ST (1956) My forty years with Ford. Norton, Baltimore

Prenting TO, Thomopoulos NT (1974) Humanism and technology in assembly line systems. Spartan-Hayden, New Jersey

Wild R (1975) Work organization, a study of manual work and mass production. Wiley, New York

Hounshell DA (1984) From the American system of mass production 1800–1932. John Hopkins University Press, Baltimore

Elsayed EA, Boucher TO (1985) Analysis and control of production systems. Prentice Hall, New Jersey

Bedworth DD, Bails J (1987) Integrated production control systems, analysis design. Wiley, New York

Chow W (1990) Assembly line design methodology and applications. CRC Press, New York

Lee MH, Rowland JJ (1995) Intelligent assembly systems. World Scientific Publishers, Singapore

Groover MP (2001) Automation, production systems and computer integrated manufacturing. Prentice Hall, New Jersey

Barkston J (2003) Henry Ford and the assembly line. Mitchell Lane Publishers, Delaware

Lambert AJD, Gupta SM (2005) Disassembly modeling for assembly, maintenance, reuse, and recycling. CRC Press, Boca Raton

Rekiek B, Delchanbre A (2010) Assembly line design. Springer, New York

York MJ (2010) Henry Ford, manufacturing mogul. Essential Library, Chicago

Grzechcr W (2011) Assembly line, theory and practice. InTech, Chicago

McGovern SM, Gupta SM (2011) The disassembly line: balancing and modeling. McGraw Hill, New York

Noble DF (2011) Forces of production. Transaction Publishers, New Jersey

Nye DE (2013) America's assembly line. MIT Press, Boston

N. T. Thomopoulos, *Assembly Line Planning and Control*,
DOI: 10.1007/978-3-319-01399-2,

Index

N. T. Thomopoulos, *Assembly Line Planning and Control*,
DOI: 10.1007/978-3-319-01399-2,

MIX
Papier aus verantwortungsvollen Quellen
Paper from responsible sources
FSC® C105338

FSC
www.fsc.org

Printed by Books on Demand, Germany